James Hamblin Smith

Elementary statics

Rvingtons mathematical series

James Hamblin Smith

Elementary statics
Rvingtons mathematical series

ISBN/EAN: 9783742833242

Manufactured in Europe, USA, Canada, Australia, Japa

Cover: Foto ©Thomas Meinert / pixelio.de

Manufactured and distributed by brebook publishing software
(www.brebook.com)

James Hamblin Smith

Elementary statics

Rivington's Mathematical Series

ELEMENTARY STATICS

Rivington's Mathematical Series.

By J. HAMBLIN SMITH, M.A.
of Gonville and Caius College, and late Lecturer at St. Peter's College, Cambridge.

A TREATISE ON ARITHMETIC. 3s. 6d.
(Copies may be had without the Answers.)

KEY TO ARITHMETIC. 9s.

ELEMENTARY ALGEBRA. Part I. 3s.
Without Answers, 2s. 6d.

KEY TO ALGEBRA. 9s.

EXERCISES ON ALGEBRA. 2s. 6d.
(Copies may be had without the Answers.)

ELEMENTS OF GEOMETRY. Containing Books 1 to 6, and portions of Books 11 and 12 of Euclid, with Exercises and Notes, arranged with the Abbreviations admitted in the Cambridge Examinations. 3s. 6d.
PART I. Containing Books 1 and 2 of Euclid, limp cloth, 1s. 6d., may be had separately.

KEY TO GEOMETRY. 8s. 6d.

ELEMENTARY STATICS. 3s.

ELEMENTARY HYDROSTATICS. 3s.

KEY TO STATICS AND HYDROSTATICS. 6s.

ELEMENTARY TRIGONOMETRY. 4s. 6d.

KEY TO TRIGONOMETRY. 7s. 6d.

BOOK OF ENUNCIATIONS. For Geometry, Algebra, Trigonometry, Statics, and Hydrostatics. 1s.

THE STUDY OF HEAT. 3s.

By E. J. GROSS, M.A.
Fellow of Gonville and Caius College, Cambridge.

ALGEBRA. Part II. 8s. 6d.

KINEMATICS AND KINETICS. 5s. 6d.

By G. RICHARDSON, M.A.
Assistant Master at Winchester College.

GEOMETRICAL CONIC SECTIONS. 4s. 6d.

Rivingtons: Waterloo Place, London

ELEMENTARY STATICS

BY

J. HAMBLIN SMITH, M.A.

OF GONVILLE AND CAIUS COLLEGE, AND LATE LECTURER AT
ST. PETER'S COLLEGE, CAMBRIDGE

NEW EDITION

RIVINGTONS

WATERLOO PLACE, LONDON

MDCCCLXXXIII

PREFACE TO THE THIRD EDITION.

THIS treatise was originally designed to explain the part of Statics required in the Previous Examination and the Second Examination for Ordinary Degrees in the University of Cambridge. It is now published in such a form, that, while serving its primary purpose, it may meet the requirements of Students in Schools, especially those who are preparing for the Local Examinations. It may also be regarded as an introduction to the works on Mechanics, which will appear in due course in RIVINGTON'S MATHEMATICAL SERIES.

The Examples have been selected from Papers set in University Examinations.

The propositions requiring a knowledge of Trigonometry are marked with *Roman* numerals.

For some explanations of the Elementary Definitions I am indebted to the late Dr. Whewell's work on " *The Philosophy of the Inductive Sciences;*" and a special acknowledgment is due from me to Dr. Parkinson, for permission to make free use of his treatise on Mechanics.

In this, as in my other publications, I have been assisted in no slight degree by Mr E. J. Gross, of Gonville and Caius College, who has taken the most lively interest in revising and correcting all that I have written.

J. HAMBLIN SMITH.

CONTENTS.

CHAPTER VIII.

CHAPTER IX.

CHAPTER X.

ELEMENTARY STATICS.

CHAPTER I.

DEFINITIONS.

1. MATTER is that, which can be perceived by the senses of sight and touch.

A BODY is *any* portion of matter.

A *RIGID* BODY is one, in which the different portions are held together in invariable positions with respect to each other.

A PARTICLE or MATERIAL POINT is a portion of matter, *indefinitely small* in all its dimensions: so that its length, breadth, and thickness are less than any assignable linear magnitude.

2. REST. When a body or particle constantly occupies the same position, it is said to be at rest.

MOTION. When the position of a body or particle is being changed continuously, it is said to be in motion.

3. FORCE. Any cause, which changes or tends to change the state of rest or motion of a body or particle, is called force.

4. STATICS is the science, which treats of the conditions, under which forces, acting on matter, produce rest.

5. LINE OF ACTION. The line of action of a force is the line, in which a particle would begin to move in consequence of the action of the force.

6. The Forces, with which we are chiefly concerned in this treatise, may be roughly divided into three classes :

(1) PRESSURES. (2) TENSIONS. (3) ATTRACTIONS.

Of the first and second kinds of force we have illustra-

pull, or lift a body, we bring into action a force acting by *pressure* or by *tension*. Imagine a gimlet to be firmly fixed in a block of wood. If we push the gimlet, we apply to the block a force acting by *pressure*. If we pull the gimlet, we apply to the block a force acting by *tension*.

Hence we obtain the following definitions:

PRESSURE. If one body be forced against another, each body is subjected to a force acting at the point of contact: such force is called pressure.

TENSION. When a body is pulled by means of a string or rod, the force exerted along the string or rod is called tension.

If we consider a string as a line of consecutive particles, when a force is applied at each end of the string, each particle of the string is pulled in opposite directions by the forces, which the consecutive particles on either side of it are compelled to exercise upon it. These forces are called *tensions*, and are the same at every particle of the string.

Suppose an engine, attached to a truck by a coupling-chain, to be just on the point of moving the truck. *Each link* of the chain is then acted upon by two equal and opposite forces, which act by means of the other links on either side of any particular link. The force, with which the part of the chain on *one side* of any particular link resists the force exerted along the chain on the *other side* of the link, is called the *tension* of the chain.

7. ATTRACTION is a force less easily conceived than pressure or tension, because it arises from the action of one body on another *at a distance from it.*

Such is the influence of the magnet on the needle: such is the influence by which the Earth attracts to itself all bodies about it: and such is the influence by which the Sun and planets attract each other.

8. All bodies fall, if unsupported, or tend to fall, if supported, towards the surface of the earth.

The direction, in which a particle would fall freely at any place, is called the *vertical line* at that place.

A plane perpendicular to this vertical line is said to be *horizontal*.

If a ball of lead be suspended at one end of a string, and we hold the other end of the string, we must exert a certain force to sustain the ball, equal to the force, with which the Earth attracts the ball. This latter force is called the *Weight* of the ball. Hence we obtain the following definition:

WEIGHT or GRAVITY is the name given to the force with which the earth attracts a body.

The tendency of bodies to the Earth results from their attraction or *Gravitation* to the Earth. This tendency is only a particular instance of the *attraction*, which is exerted by every body upon those about it; and this attraction of one body to another arises from the attraction of every particle of matter to every other, which is called *Universal Gravitation*.

9. DENSITY. The Density of a substance is the degree of closeness, with which the particles composing the substance are packed together.

10. VOLUME is the amount of space occupied by a body.

The Volume, Bulk, or Solid Content of a body is measured by the number of times a certain cubical unit must be repeated, to fill up the space occupied by the body.

Thus, when we say that the volume of a body is 8 cubic inches, we mean that a cubical unit, which we call a cubic inch, must be repeated 8 times, to fill up the space occupied by one body.

11. EQUILIBRIUM. If several forces acting on a particle, or on a body, are so related, that no motion of the particle or the body takes place, the forces are said to be in equilibrium.

Two forces, which, acting in opposite directions, keep each other in equilibrium, are necessarily and manifestly equal. If we see two boys pulling at two ends of a rope, so that neither of them in the smallest degree prevails over the other, we have a case in which *two* forces are in equilibrium. If three hooks be fixed in a log of wood, two at one end and one at the other, and if the efforts of two boys pulling at ropes, attached to the two hooks at one end, be just counteracted by the effort of a man pulling at a rope, attached to the hook at the other end, we have a case in which *three* forces are in equilibrium: and this illustration may be extended to *four, five* or more forces.

Again, if a number of rings be inserted round a block of wood, if a rope be attached to each ring, and a boy set to pull at each rope, it is easy to conceive such a disposition of the forces exerted by the boys, that no motion of the block may take place. Here then we have a case, in which a number of forces, not acting in parallel directions, are in equilibrium.

12. When two men pull at a rope in the same direction, we know that the force, which they exert, is equal to the sum of the forces, which they would separately exert. When two stones are put in a basket suspended by a string, their weights are added and the sum is supported by the string. Thus we see that forces, acting together in the *same* direction, may be added together to obtain their combined effect.

Since two opposite forces, which balance each other, are equal, each force is measured by that which it balances; and since forces are capable of addition, a force of any magnitude is measured by adding together a proper number of such equal forces.

Thus a heavy body, which, appended to an elastic spring, will draw it through one inch, may be taken as the *unit of weight*. Then, if we remove this body, and find a second heavy body, which will also draw the spring through one inch, this second body is also a unit of weight. In like manner we might go on to a third and a fourth equal body; and adding together the two, or the three, or the four heavy bodies, we have a force twice, or three times, or four times the unit of weight. And with such a collection of heavy bodies, or *weights,* we can readily measure all other forces; for, since forces that keep a body at rest must be equal in their opposite effect, we conclude that *any statical force is measured by the weight which it will support.*

13. To *measure* forces we fix upon some definite force for our standard, or *unit,* and then any other force is measured by the number of times it contains this unit, and this number is called the *measure* of the force.

It is usual to take as the unit of force that force which will sustain, when acting vertically upwards, a weight of one pound avoirdupois. The measures of forces, which will sustain 1 lb., 2 lbs., 3 lbs., P lbs., will then be 1, 2, 3, P respectively.

14. Two forces are *commensurable* when a force can be

taken as the standard of measurement, such that it is contained in each an exact number of times.

15. *Method of estimating forces.*

The three elements specifying a force, all of which must be known in order to estimate the effect of the force, are

(1) The point of application of the force.

(2) The direction in which the force acts.

(3) The magnitude of the force.

16. *Method of representing forces.*

Forces may be represented by straight lines: for

(1) A straight line can be drawn from any point, and thus it will represent a force with respect to the point of application.

(2) A straight line can be drawn in any direction, and thus it will represent the direction of a force.

(3) A straight line can be drawn of such a length, as to contain as many units of length as the given force contains units of force, and thus it will represent the magnitude of a force.

Thus, suppose we are speaking of a force of 5 lbs., acting at the middle point of a horizontal rod, and inclined at an angle of 45° to the horizon.

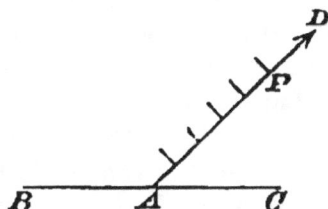

Let *BC* represent the rod, *A* the middle point of the rod.

Draw *AD* making an angle of 45° with *AC*.

Mark off a portion of the line *AD*, suppose *AP*, containing 5 units of length, that is, as many units of length as there are units of force in the given force.

Then we may say that *AP* represents the given force in

(1) In point of application, at A the middle point of the rod.

(2) In direction, as being inclined at an angle of 45° to the horizon.

(3) In magnitude, by the number of units in its length.

Next, suppose that we have to represent two forces of 5 lbs. and 7 lbs., applied to a point in directions at right angles to each other.

Taking any line we please to represent the unit of length, we draw two lines AB, AC at right angles to each other,

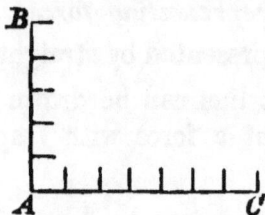

the one containing our unit of length 5 times, and the other containing it 7 times, and the lines AB, AC will properly represent the two forces acting at the point A.

17. *On the Transmissibility of Force.*

It is plain that two equal and opposite forces, P, Q, applied

at the extremities of a straight rigid rod AB, and acting in direction of the rod, will be in equilibrium.

This result will be true whatever be the length of the rod: and hence we infer that P will balance Q, at whatever point of the rod Q be applied; in other words, the effect of Q is the same, at whatever point of the rod, whether at B, C, or any other point, it may be applied, the direction remaining the same.

Suppose a piece of wood of any shape, say like a horse-shoe, to be laid on a *smooth* table, and to be acted on by a force, A, B, C, D being any four points in the line of action of the force, and in the wood. Then the only force tending to move the body is the aforesaid force, and it

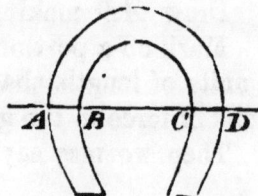

is found that the body will move in precisely the same manner, whether the force be applied at *A*, *B*, *C*, or *D*.

These considerations lead us to the following principle, called the *principle of the transmissibility of force.*

THE EFFECT OF A FORCE ON A RIGID BODY, TO WHICH IT IS APPLIED, WILL BE THE SAME, IF WE SUPPOSE THE FORCE TO BE APPLIED AT ANY POINT IN THE LINE OF ACTION, PROVIDED THE POINT BE RIGIDLY CONNECTED WITH THE BODY.

The following is another example of this principle. If *ABC* be a chain fastened at *A* to a hook fixed in a beam, and a weight *W* be suspended from *C*, then the pressure on the hook is *W* + weight of chain *AC*. If now we remove the part of the chain below *B*, and suspend *W* from *B*, the pressure on the hook is *W* + weight of chain *AB*.

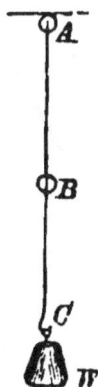

Thus the pressure on the hook caused by *W* acting at *C* exceeds the pressure caused by *W* acting at *B* by the weight of the chain *BC*, and *W*, so far as *its own effect is concerned, produces the same pressure on* A, *whether it be applied at* C *or* B.

NOTE. It may here be remarked, that if a weight W be

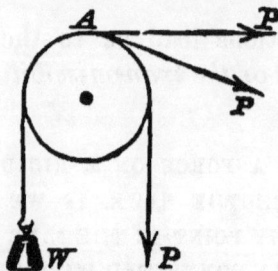

attached to one end of a fine string, and the string be made to pass over a small grooved wheel, fixed in position and moveable about its centre, it is proved by experiment that the force P, which has to be applied at the other end of the string to keep W at rest, is the same, whatever may be the angle between the two parts of the string WA and PA.

The same is true with regard to the tension of a string that passes over a smooth peg. Thus, if a picture frame be suspended symmetrically by a cord passing over a nail A, the tension of the portions of the cord AB and AC is the same.

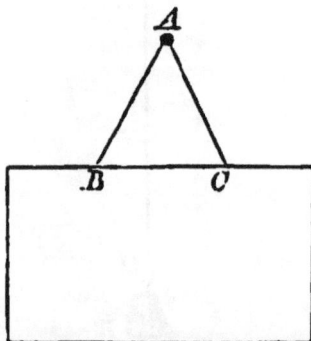

CHAPTER II.

ON COMPONENT AND RESULTANT FORCES.

18. Suppose O to be a material particle, and AB a straight line of unlimited length passing through O.

$$A \quad Q \quad N \quad O \quad M \quad P \quad B$$

Let a force P act on O in the direction AB,
and............... Q ....BA.

Set off in OB the line OM, containing as many units of length as P contains units of force; and in OA set off ON, containing as many units of length as Q contains units of force: then OM and ON will represent P and Q in the point of application, in direction and in magnitude.

The learner must be careful to speak of a line not *as a force*, but *as representing a force*. Thus we say here that OM *represents* the force P. The order, in which the letters O and M stand, serves to indicate the direction, in which the force P acts: OM representing a force that acts from O towards M, whereas MO would represent a force of equal magnitude acting from M towards O.

19. Suppose O to be a material particle, and AB a straight line of unlimited length passing through O.

$$A \quad O \quad P \quad Q \quad B$$

If two forces, P and Q, act simultaneously on O in the *same* direction OB, they are equivalent to a single force $P+Q$ acting along OB.

20. If the forces P and Q act in *opposite* directions, OB

and OA respectively, they are equivalent to a single force equal in magnitude to the *difference* between P and Q, and acting in the direction of the greater of the two forces. And, generally, a system of forces, acting in the same line AB on a particle O, is equivalent to a single force, equal in magnitude to the excess of the sum of the forces which act in one direction, over the sum of those which act in the opposite direction, and acting in the direction of the greater sum.

21. If P and Q be two forces acting simultaneously in the lines AB, AC, *not in the same straight line*, on a free particle A, the particle will begin to move in some definite direction

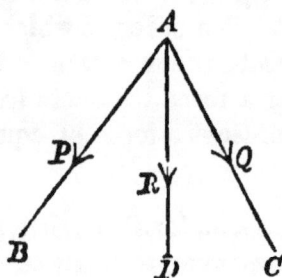

AD within the angle BAC, which is less than two right angles. The particle will move continuously along AD further and further the longer P and Q continue to act in their original directions.

Now suppose P and Q to be removed and a single force R to act on the particle in the direction AD, so as to cause the particle to move over the same space in any given interval of time, as it would have moved over in that interval had it been acted on by P and Q. The force R is called the RESULTANT of the forces P and Q, and we obtain the following definition:

DEF. A force, whose effect is equivalent to the combined effects of two or more other forces, is called their *Resultant*, and, with reference to this resultant, each of the other forces is called a *Component*.

NOTE.   In Arts. 19 and 20 we treated of a single force equivalent to a system of forces acting *in the same straight line*, to which force the term *Resultant* may, in accordance with the definition just given, be applied.

22.   If P and Q be *equal* forces, the particle A will move in a straight line *bisecting* the angle BAC; for there is no reason why it should move more towards the direction of one force, than towards the direction of the other.   Hence the line of action of the resultant of two equal forces bisects the angle between their lines of action.

23.   Now suppose a particle A to be kept at rest by three forces P, Q, S.   Then S may be regarded as neutralizing the joint effect of P and Q.

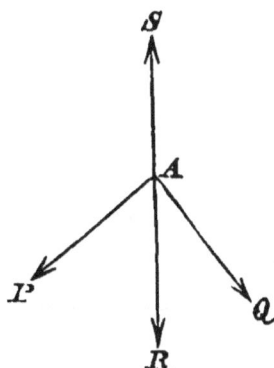

Hence, if R be the resultant of P and Q, R and S must be equal and opposite forces, for since R produces the same effect as that which is produced by the combined action of P and Q, R and S are also capable of keeping the particle A at rest, and this they cannot do, unless they be equal in magnitude and opposite in direction.

24.   *Illustrations of Component and Resultant Forces.*

Since a clear conception of the meaning of the terms *Component* and *Resultant* is necessary for a right understanding of Statical principles, we shall give in this Article two rough illustrations, which may serve to explain the definition given in Art. 21.

A is a block of stone to be drawn along a level road RD.

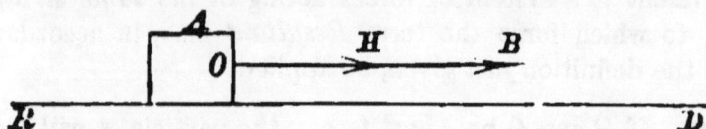

Suppose two horses to be attached to the stone at O, in such a manner, that each exerts a force along the line OHB, one at H and the other at B.

Now suppose the horses to be removed, and a traction-engine to be applied to the block at O, so as to move it in the direction OHB with the same drawing-power as that of the two horses.

The horses will represent the Component Forces.

The traction-engine will represent the Resultant.

CA, MN are the straight and parallel edges of the banks of a river.

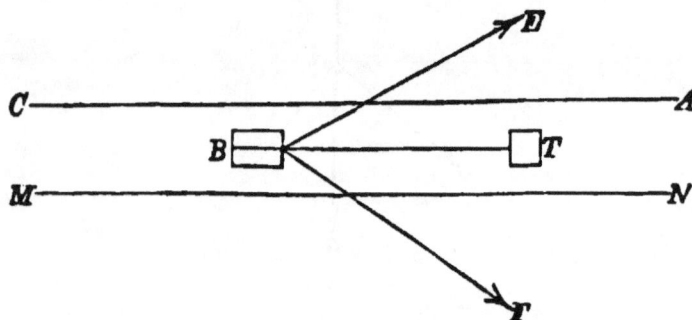

B is a barge in the middle of the stream.

Suppose two horses of equal power to be pulling at ropes attached to the same point of the barge, the ropes being inclined at equal angles to the line BT which passes along the middle of the stream, parallel to the banks.

Then the barge will move along the line BT.

Now suppose the horses to be removed, and a steam-tug to be attached to the barge, so as to move it in the direction BT, just in the same manner as it moved when the horses were pulling it.

The horses will represent the Component Forces.

The steam-tug will represent the Resultant.

25.   From the two illustrations of Component and Resultant Forces, which have been given, we may derive examples
of forces in Equilibrium.

For, first, suppose that while the horses are pulling at
the stone the traction-engine is applied *to the opposite side* of
the stone, so as to pull the stone with the same drawing-power
as that of the horses.

Then the force exerted by the engine will counteract the
forces exerted by the horses, and the three forces will be
in equilibrium.

Then also it is plain that when three forces are in
equilibrium, one of them is equal and opposite to the resultant of the other two.

Precisely the same results will follow, if in the second
illustration we suppose the steam-tug to be applied *to the
opposite end* of the barge.

26.   Let P, Q be forces acting at a point A, and P', Q'
be other forces acting at a point B rigidly connected with
A.   If P', Q' produce the same effect as that which is

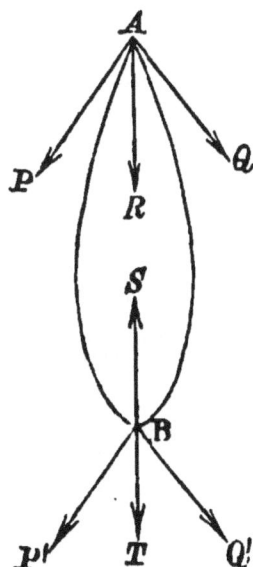

produced by P, Q, then the resultant of P, Q lies in the
straight line joining A and B.   For suppose the resultant of
P', Q' to be in the direction BT.   Then the force counter-
acting the effect of P', Q' will lie in the line BS opposite

to BT. Now this force is to counteract the effect of P and Q, and therefore it must be in the same line with AR, the direction of the resultant of P, Q. (Art. 23.) Hence the resultant of P, Q passes through B.

27.   Before proceeding further, we will state three axioms which are the groundwork of much that follows:

Axiom I.   A force may be supposed to act at any point in the line of its direction.   Art. 17.

Axiom II.   Forces may have equivalent forces substituted for them.

Axiom III.   When two or more forces are in equilibrium, their resultant is *zero*.

CHAPTER III.

THE PARALLELOGRAM OF FORCES.

26. WE proceed to establish an important theorem, which enables us to determine the resultant of any two forces acting at a point. The theorem is called the Parallelogram of Forces, and may be thus enunciated.

If two forces, acting at a point, be represented in magnitude and direction by two straight lines drawn from that point, and if a parallelogram be constructed, having these two lines for adjacent sides, then that diagonal of the parallelogram, which passes through the point of application of the forces, will represent their resultant in magnitude and direction.

That is, if the two forces P, Q be represented by AB, AC, and the parallelogram $ABDC$ be completed, their resultant R will be represented by the diagonal AD.

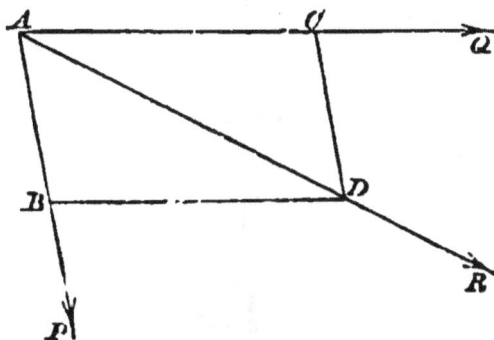

In other words, if AB and AC contain as many units of length as P and Q contain units of force, the resultant, R, of P and Q will act in the line AD, and will contain as many units of force as AD contains units of length.

29. The truth of this theorem is illustrated by the following experiment.

Two small pullies M and N are attached to a vertical wall, and a string is passed over them, having weights of 3 lbs. and 4 lbs. attached to its ends. A weight of 5 lbs. is then suspended from a point A of the part of string between the pullies: this will draw the string down so as to form an angle MAN, and the apparatus will settle itself in a state of rest as represented in the diagram.

The tensions of the strings AM, AN, are equivalent to 3 lbs. and 4 lbs., and their resultant is equal and opposite to the weight 5 lbs.

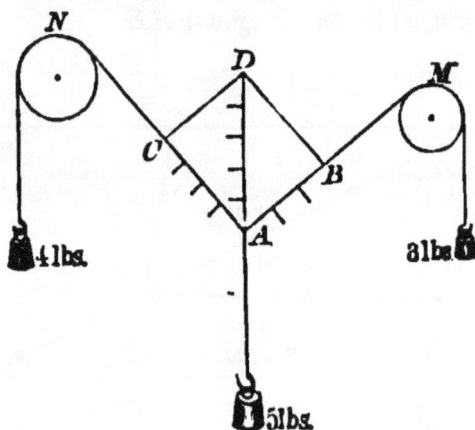

Now mark off a distance AB, along the string AM, containing three units of length, and a distance AC, in the direction of the string AN, containing 4 units of length, and complete the parallelogram $ABDC$. Then it will be found that the diagonal AD is a vertical line, and therefore in the *direction* of the line in which the weight of 5 lbs. acts, and if AD be measured, it will be found to contain 5 units of length, and therefore it represents the weight of 5 lbs. in *magnitude*. This shows that AD represents in direction and magnitude the resultant of the forces represented in direction and magnitude by AB, AC.

30. If AB and AC, the sides of a rhombus $ABDC$, represent in magnitude and direction the equal forces P and Q, which act at the point A, then must the diagonal AD represent in direction R, the resultant of P and Q.

For $\because AB = AC$, and AD is common, and $BD = CD$,

$$\therefore \angle BAD = \angle CAD, \qquad \text{(Eucl. I. 8.)}$$

$\therefore AD$ bisects the angle between the directions of P and Q.

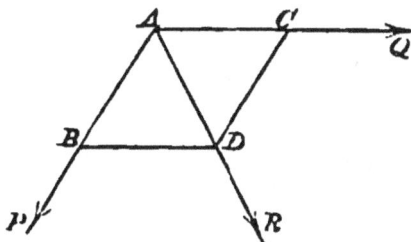

But the direction of R bisects the angle between the directions of P and Q. (Art. 22.)

$$\therefore AD \text{ represents } R \text{ in direction.}$$

Now, by the principle of the transmissibility of force, R, acting along a rigid rod AD, will produce the same effect when applied at D, as that which it produces when applied at A. Hence for P and Q acting at A we may substitute R, acting in the direction AD, at D.

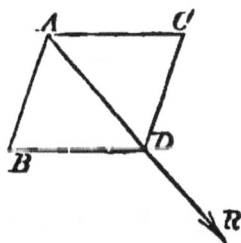

Again, R acting at D may be replaced by two forces P and Q, acting in directions parallel to AB and AC respectively, that is, acting in the directions CD and BD, thus:

Hence if $ABDC$ be a rhombus, whose sides are rigidly connected, two *equal* forces P and Q will have the same effect on a particle A, whether they be applied along AB and AC, or along CD and BD.

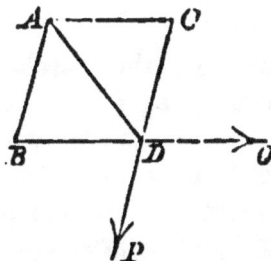

31. We now proceed to the Mathematical Proof of the Parallelogram of Forces, which is divided into three Parts.

Part I. To prove that the resultant acts in the direction of the diagonal, when the component forces are *commensurable.*

First, to shew that the proposition is true for forces P and P. When the component forces are *equal* their resultant bisects the angle between the directions of the forces, and therefore acts along the diagonal. Thus the proposition is true for P and P.

Next, to shew that the proposition is true for forces P and $2P$.

Let P, Q, R be three equal forces.

Let P act at A in the direction AB, let Q and R act at A in the direction ACE. Take AB and AC to represent P and Q in magnitude, and since R may be supposed to act at any point in the line ACE, which is rigidly connected with A (Ax. I.), let R act at C, and take CE to represent R in magnitude.

Complete the parallelograms BC, DE, and draw the diagonals AD, CF. The resultant of P and Q acts along AD. Let P and Q be replaced by this resultant (Ax. II.) and let it act at D. Then for this resultant acting at D we may substitute the two forces P and Q, acting in the lines CD, DF, which are respectively parallel to AB, AC.

Now suppose P to act at C, and Q to act at F. (Ax. I.)

Then P and R, acting at C, have a resultant acting along CF: let them be replaced by this resultant, and let it act at F.

For this resultant we may substitute the forces P and R, acting at F in the lines EF and DF. (Ax. II.)

Thus we have shewn that the forces P and $Q+R$ which are applied at A, may be supposed to act at F without altering their combined effect;

∴ F is a point in the direction of the resultant of P and $Q+R$ (Art. 25),

∴ AF is the direction of the resultant of P and $Q+R$,
that is, of P and $2P$.

By a similar process we can shew that the proposition is true for P and $3P$, using the annexed diagram.

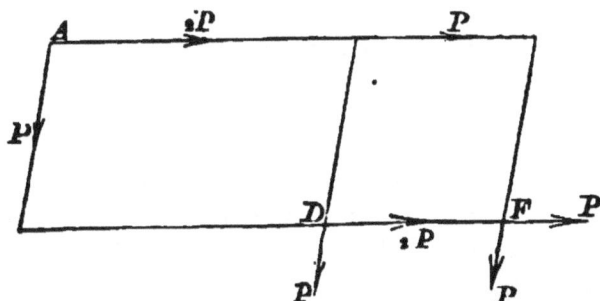

Similarly, it may be shewn to be true for P and $4P$, for P and $5P$, and so for P and mP, m being any whole number.

Now since the proposition is true for mP and P, it may be shewn to be true for mP and $2P$, by using the annexed figure.

So also it may be shewn to be true for mP and $3P$, for mP and $4P$, and so for mP and nP, n being any whole number.

Now any two commensurable forces may, by assigning a proper value to P, be expressed by mP and nP.

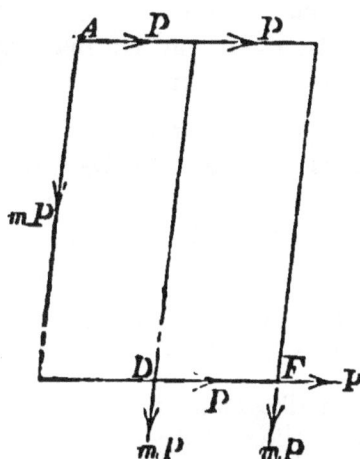

Hence Part I. is proved.

32. Part II. To prove that the resultant acts in the direction of the diagonal, if the forces are *incommensurable*.

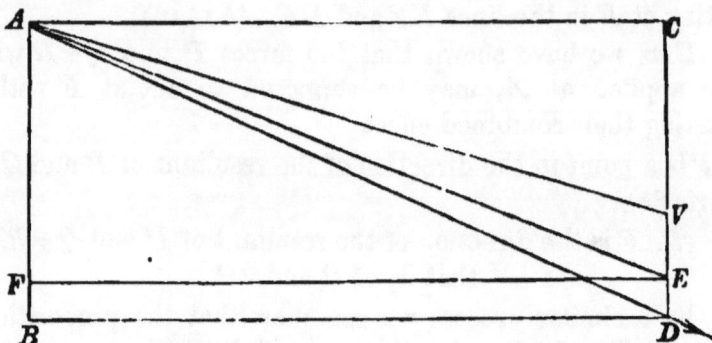

Let AB, AC represent two incommensurable forces.

Complete the parallelogram $ABDC$, and if AD be not the direction of the resultant, let it be some other line, as AV.

Let AC be divided into an integral number of equal parts, each less than DV, which is always possible, and mark off from CD portions equal to these, the last division E clearly falling between D and V.

Complete the parallelogram CF by drawing EF parallel to AC.

Then AC, AF represent *commensurable* forces, and the resultant of the forces represented by AC, AF will be in the direction AE, and we may suppose this resultant to be substituted for them.

The resultant then of the forces represented by AC and AB is equivalent to the resultant of two forces, one acting in the direction AE, the other represented by FB, and which may therefore be supposed to act at A in the direction AB; and this resultant must lie *within* the angle BAE.

But, by hypothesis, it acts in the direction AV, *without* the same angle, which is absurd.

In like manner it may be shewn that no direction but AD can be that of the resultant of the forces represented by AB, AC.

Thus the theorem has been proved, so far as *the direction of the resultant* is concerned.

33. Part III. *To prove that the diagonal represents the resultant in magnitude.*

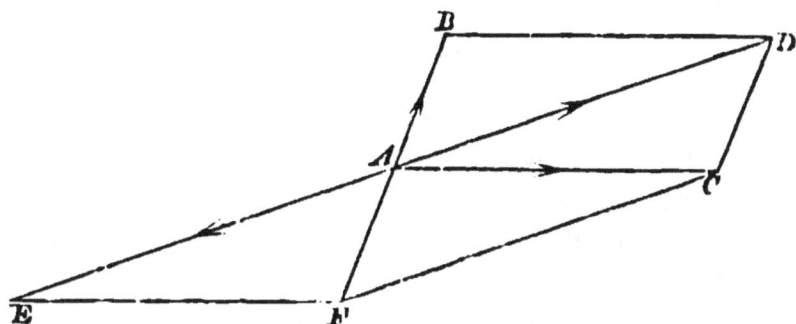

Let AB, AC represent the component forces in magnitude and direction.

Complete the parallelogram $ABDC$: join AD.

In DA produced take AE of such a length, as to represent the magnitude of the resultant of the forces represented by AB, AC.

Complete the parallelogram $AEFC$: join AF.

Now AB, AC, AE represent three forces which are in equilibrium.

Therefore AB represents a force equal *and opposite* to the resultant of the forces represented by AC, AE (Art. 23).

But the resultant of the forces represented by AC, AE *lies in the direction of AF.*

Therefore AB is in the same straight line with AF.

Therefore $AFCD$ is a parallelogram;

and $\therefore AD = FC$;

but $FC = AE$;

                              $\therefore AD = AE$;

$\therefore AD$ represents in magnitude the resultant of the forces

34.   When we determine, by means of the Parallelogram of
Forces, the single force, which is equivalent in its effect to
the joint effect of two other forces, we are said to *compound*
those forces.

We shall now give some simple examples on the composi-
tion of forces, so far as the method can be illustrated by easy
Geometrical processes.   The following Theorems are of frequent use:

1.   The diagonals of a Square bisect the angles.
2.   The diagonals of a Rhombus bisect the angles.
3.   The diagonals of a Parallelogram bisect each other.
4.   The perpendicular, dropped from the vertex of an
equilateral triangle on the base, bisects the base, and also the
vertical angle.

Consequently, if the angles, A, B, C, of a triangle be $90°$
$60°$ and $30°$ respectively, then will the side BC be double of
the side BA.

35.   *If two forces act at the same point, in directions at
right angles to each other, to find the magnitude and direction of their resultant.*

Let AC, AB represent two forces
P, Q, acting at right angles to each
other at the point A.

Complete the rectangular parallelogram $ABDC$.

Then the diagonal AD will represent
R, the resultant of P, Q.

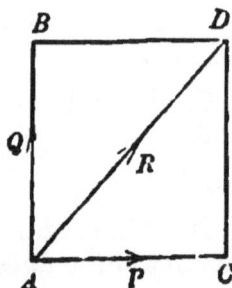

Now, since the angle DCA is a right angle,

$$AD^2 = AC^2 + CD^2;$$
$$\therefore AD^2 = AC^2 + AB^2;$$
$$\therefore R^2 = P^2 + Q^2;$$
$$\therefore R = \sqrt{P^2 + Q^2};$$

and thus we obtain the *magnitude* of the resultant

The *direction* of the resultant is known, if we know the
size of the angle DAC.

For certain simple relations between the sides of the triangle ADC, we can determine the angle DAC by *geometry*. Thus, if AB and AC represent equal forces, AC and CD are equal, and DAC is half a right angle.

EXAMPLES WORKED OUT.

1. *Two forces of 12 lbs. and 16 lbs. act on a particle in directions at right angles to each other: find the magnitude of their resultant.*

Let AB and AC represent the component forces.

Complete the rectangle $ABDC$.

Then AD represents the resultant.

Let the measure of AD be x.

Then
$$x^2 = (16)^2 + (12)^2 ;$$
$$\therefore x^2 = 256 + 144 ;$$
$$\therefore x^2 = 400 ;$$
$$\therefore x = 20.$$

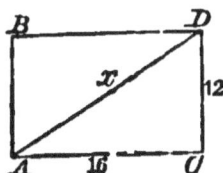

Hence the magnitude of the resultant is 20 lbs.

2. *Two forces of 8 lbs. each act at an angle of 60° on a particle: find the magnitude and direction of the resultant.*

Let AB, AC represent the component forces.

Complete the rhombus $ABDC$, and draw DE at right angles to AC produced.

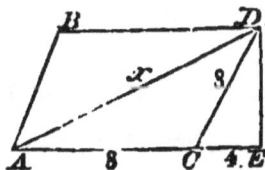

Then $\because \angle DCE = \angle BAC = 60°$,

and $\angle CED = 90°$,

$\therefore \angle CDE = 30°;$

and $\therefore CD = 2CE.$ (Art. 34)

Let x be the measure of AD.

Then $\because AD^2 = AC^2 + CD^2 + 2AC \cdot CE;$ (Eucl. II. 12)
$$\therefore x^2 = 8^2 + 8^2 + 2 \times 8 \times 4 ;$$
$$\therefore x^2 = 64 + 64 + 64 ;$$
$$\therefore x^2 = 64 \times 3 ;$$
$$\therefore x = 8\sqrt{3}.$$

Hence the *magnitude* of the resultant is $8\sqrt{3}$ lbs.

The *direction* of the resultant can also be determined. For since the diagonal of the rhombus $ABDC$ bisects the angle BAC, we know that the resultant makes an angle of $30°$ with each of the components.

NOTE. By the use of Trigonometry we are enabled to work examples of this kind, by means of a general formula, as we now proceed to shew.

xxxvi. *If two forces act at the same point, and the angle between their lines of direction is given, to find expressions for the magnitude and direction of their resultant.*

Let AC, AB represent two forces P, Q, acting upon the point A, and let a be the angle between their lines of direction.

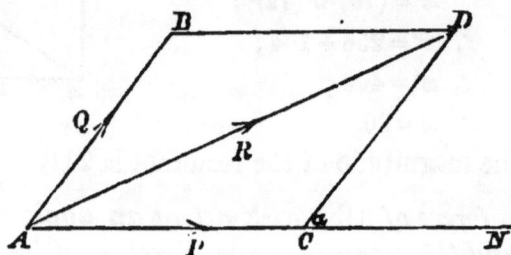

Complete the parallelogram $ABDC$, and produce AC to N.

Then the angle $DCN = a$. Join AD.

Then AD will represent R, the resultant of P, Q.

Now we know by Trigonometry (Art. 179) that

$$AD^2 = AC^2 + CD^2 - 2AC \cdot CD \cdot \cos ACD;$$

also (Trig. Art. 101), $\cos ACD = -\cos DCN$

$$= -\cos a;$$

$$\therefore AD^2 = AC^2 + AB^2 + 2AC \cdot AB \cdot \cos a;$$

$$\therefore R^2 = P^2 + Q^2 + 2PQ \cdot \cos a;$$

and thus we obtain an expression for the *magnitude* of the resultant.

The *direction* of the resultant is known, if we know the size of the angle DAC.

Let $\angle DAC = \theta$; then $\angle ADC = a - \theta$.

Then $\dfrac{\sin \theta}{Q} = \dfrac{\sin (a-\theta)}{P}$. (Trig. Art. 178.)

$\therefore P . \sin \theta = Q . \sin a \cos \theta - Q . \cos a . \sin \theta$;

$\therefore \sin \theta . (P + Q . \cos a) = Q . \sin a \cos \theta$;

$\therefore \tan \theta = \dfrac{Q . \sin a}{P + Q . \cos a}$.

EXAMPLES.—I.

1. Three forces of 3 lbs., 4 lbs. and 5 lbs., respectively, act on a body, their directions being all in the same straight line: find a fourth force which will balance them.

2. Find the resultants of the forces 5 lbs., 11 lbs., 13 lbs., according to the different possible arrangements of them;— all three acting in the same straight line.

3. P and Q are two forces applied to a particle in directions at right angles to one another; P is 90 lbs., Q is 120 lbs.; find the magnitude of the resultant.

4. Forces of 36 lbs. and 48 lbs. act on a particle in directions at right angles to one another: find the magnitude of the resultant.

5. Three forces, whose magnitudes are 6, 8 and 10 lbs. respectively, acting upon a point, keep it at rest: prove that the directions of two of the forces are at right angles to each other.

6. Place three forces, which are in the ratio of 3, 4 and 5, so that they may keep a particle at rest.

7. If two forces, acting at right angles to each other, be in the ratio of $1 : \sqrt{3}$, and their resultant be 10 lbs., find the forces.

8. Forces of 3 lbs. and 5 lbs. act on a point at an angle of 30°. Find the magnitude of the resultant.

9. Forces of 10 lbs. and 7 lbs. act on a point at an angle of 60°. Find the magnitude of the resultant.

10. Forces of 9 lbs. and 11 lbs. act on a point at an angle of 135°. Find the magnitude of the resultant.

11. Two forces of 4 lbs. and 5 lbs. are inclined to one another at an angle of 45°, determine the magnitude of their resultant.

12. Two equal forces act upon a point. If the angle between their directions be 60°, find the resultant.

13. Two forces act at a point. Shew that the forces are equal, if, when the direction of one of the forces is reversed, the direction of their resultant is at right angles to the direction of their resultant before the change.

14. A weight of 10 lbs. is suspended by a string AB from the fixed point A. A force F acts horizontally at B on the string. What must be the magnitude of F in order that the angle ABF may be 120°?

CHAPTER IV.

ON THE TRIANGLE AND POLYGON OF FORCES.

37.   THE TRIANGLE OF FORCES.

If three forces, acting at a point, can be represented in magnitude and direction by the sides of a triangle, taken in order, they will be in equilibrium.

Let AB, BC, CA, the sides of the triangle ABC, taken in order, represent in magnitude and direction three forces, P, Q, R, acting at the point O.

Complete the parallelogram $ABCD$.

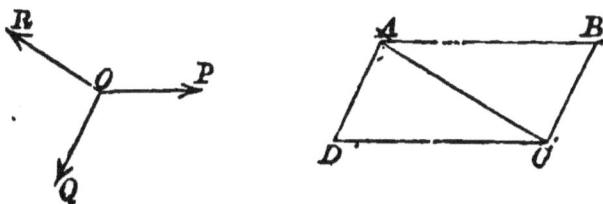

Then, since AD is equal and parallel to BC, the force represented in magnitude and direction by BC will also be represented in magnitude and direction by AD.

Therefore the forces P, Q will be represented in magnitude and direction by AB, AD.

Now AC represents the resultant of two forces represented by AB, AD.

Hence AC represents in magnitude and direction the combined effect of P and Q.

Therefore AC, CA represent in magnitude and direction the combined effect of P, Q, R.

But forces represented by AC, CA will clearly be in equilibrium.

Therefore P, Q, R will be in equilibrium.

NOTE. The converse of this proposition is true, that is, if three forces, acting at a point be in equilibrium, they can be represented in magnitude and direction by the sides of a triangle, taken in order. This is a particular case of a more general theorem, which we now proceed to prove.

38. *If three forces, acting at a point, be in equilibrium, and any triangle be constructed, having its sides parallel to the directions of the forces, the sides of the triangle shall be proportional to the forces.*

Let P, Q, R be three forces, which, acting at the point A, are in equilibrium.

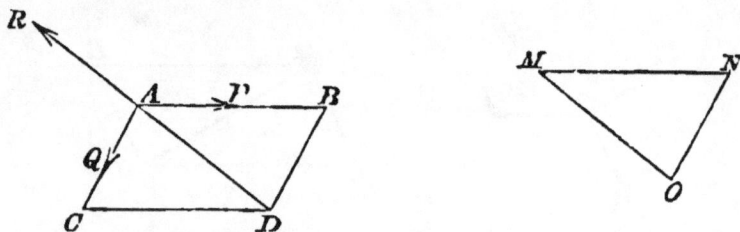

Let AB, AC represent P and Q.

Then DA, the diagonal of the parallelogram $ACDB$, will represent R.

Now construct a triangle MNO, whose sides are parallel to the sides of the triangle ABD.

Then ABD, MNO are similar triangles.

Hence $\qquad MN : NO = AB : BD \qquad$ (Eucl. VI. 4.)

$$= P : Q$$

and $\qquad NO : OM = BD : DA$

$$= Q : R$$

and $\qquad OM : MN = DA : AB$

$$= R : P.$$

COR. If two forces P and Q act at a point, and have a resultant R', and if a triangle MNO be constructed, whose sides are parallel to P, Q, R', then must

$$MN : NO : OM = P : Q : R'.$$

For let R be the force, which would be in equilibrium with P and Q : then the direction of R is opposite to that of R', and therefore the sides of the triangle MNO are parallel to P, Q, R, which are in equilibrium :

$$\therefore MN : NO : OM = P : Q : R;$$

and the magnitude of R' = magnitude of R,

$$\therefore MN : NO : OM = P : Q : R'.$$

xxxix. *If three forces, acting at a point, be in equilibrium, each force is proportional to the sine of the angle contained between the directions of the other two.*

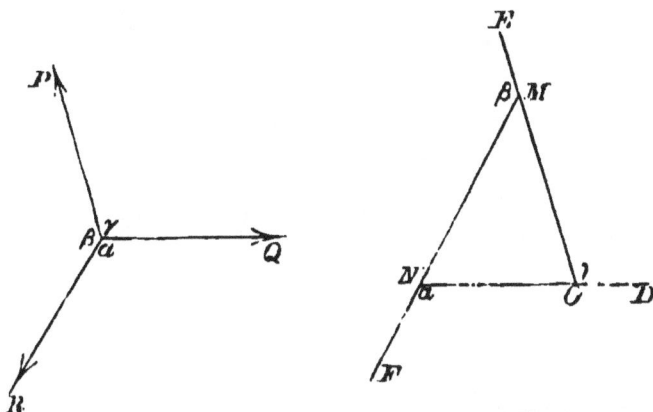

Let P, Q, R be the three forces :

α, β, γ the angles between the lines of direction of the forces.

Construct a triangle MNO, whose sides MO, ON, NM are parallel, and therefore proportional, to P, Q, R.

Produce the sides to D, E, F.

Then the exterior angles ONF, NME, MOD are equal to α, β, γ respectively.

Now

$$P : Q : R = MO : ON : NM$$

$$= \sin ONM : \sin NMO : \sin MON, \text{(Trig. Art. 178)},$$

$$= \sin ONF : \sin NME : \sin MOD, \text{(Trig. Art. 101)},$$

$$= \sin α : \sin β : \sin γ.$$

40. THE POLYGON OF FORCES.

If any number of forces, acting at a point, can be represented in magnitude and direction by the sides of a polygon taken in order, they will be in equilibrium.

Let any number of forces P, Q, R, S, T, acting at the point O, be represented in magnitude and direction by the sides of the polygon $ABCDE$, taken in order, thus, AB, BC, CD, DE, EA.

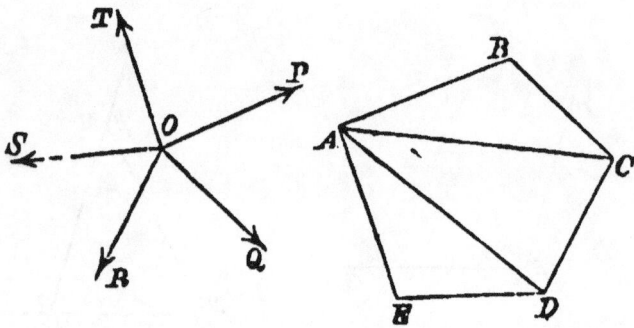

Join AC, AD.

Now AB, BC represent P, Q in magnitude and direction;

$\therefore AC$ represents the joint effect of P, Q ....

$\therefore AC$, CD represent the joint effect of P, Q, R ....

$\therefore AD$ represents...

$\therefore AD$, DE represent the joint effect of P, Q, R, S ....

$\therefore AE$ represents...

$\therefore AE$, EA represent the joint effect of P, Q, R, S, T............

Now forces represented by AE, EA will clearly be in equilibrium;

$\therefore P$, Q, R, S, T will be in equilibrium.

NOTE. The converse of this proposition is true, that is, if a number of forces, acting at a point, be in equilibrium, they can be represented in magnitude and direction by the sides of a polygon, taken in order. But it is not true that if the sides of *any* polygon, taken in order, represent the forces in direction they will also represent them in magnitude, because the sides about the equal angles of equiangular polygons are not necessarily proportional.

EXAMPLES.—II.

The following are Examples to illustrate the principles explained in this and the preceding Chapters.

1. If two equal forces P and P, acting at an angle of 60°, have the same resultant as two equal forces Q and Q, acting at right angles, shew that $P : Q = \sqrt{2} : \sqrt{3}$.

2. Two forces make an angle 120° with each other. If a third force equal to one of them form with them a system in equilibrium, compare the magnitudes of the forces.

3. (1) $ABCD$ is any quadrilateral, and E, F are the middle points of BC, AD. Prove that the system of forces represented by AE, DE, BF, CF is in equilibrium.

   (2) If two forces are represented in magnitude and direction by a side, length S, and diagonal of a square: shew that the square of their resultant will be equal to $5S^2$.

4. The direction of each of two equal forces makes with that of a third force an angle of 150°. If the three forces be in equilibrium, compare their magnitudes.

5. (1) Forces are represented by the sides AB, AC of a triangle ABC. If their resultant passes through the centre of the circle described about the triangle ABC, prove that the triangle is either isosceles or right-angled.

   (2) If three forces in arithmetical progression keep a particle at rest when the least acts at right angles to the least but one, prove that the common difference is one-third of the least force.

6. Two forces, of 2 lbs. and 3 lbs. respectively, act at a point, their directions making an angle of 60° with each other. Find the magnitude of their resultant.

7. There are two forces acting at a point making an angle of 60° with each other: the resultant is a force of 3 lbs., and one of the component forces is 2 lbs.: find the other.

8. Two forces, P and $\sqrt{2}.P$, act upon a particle. P acts towards the West, $\sqrt{2}.P$ towards the North-East. Find the direction and magnitude of their resultant.

9. Three forces P, $\sqrt{3}.P$, and $2P$ act on a particle. What must be the angles between their respective directions, in order that there may be equilibrium?

10. Three forces, P, Q, R, acting upon a particle, keep it in equilibrium. P acts towards the North, Q towards the East, and R towards the South-West. Find the ratios between the forces.

11. If three equal forces, acting upon a particle, keep it at rest, shew that their directions must be equally inclined to each other.

12. If the component forces be inclined at 120°, and the resultant be perpendicular to one of them, compare the forces.

13. If the forces be P and $2P$, and the angle between them four-thirds of a right angle, determine the magnitude of the resultant.

14. If two forces, acting at right angles to each other, have a resultant, which is double the smaller of the two forces, find its direction.

15. If two forces be inclined to each other at an angle of 135°, find the ratio between them, when the resultant is equal to the smaller force.

16. Two strings, at right angles to each other, support a weight, and one string makes an angle of 30° with the vertical line. Compare the tensions of the strings.

17. What will be the direction of the resultant, (1) if one of the components be twice as great as the other, and the angle between their directions 120 degrees, (2) if the components be equal, and one act due East, and the other North-West?

18. If AB, AC represent two forces, and D be the middle point of BC, then the resultant will act along AD, and its magnitude will be represented by $2AD$.

19. If three forces keep a point at rest, prove that the angle between the two greatest is larger than the angle between any other two.

20. Show that if the angle, at which two forces are inclined to each other, be increased, their resultant is diminished.

21. The ends of a string are tied to the rings of a picture, and the string is then passed over a nail, from which the picture hangs. Shew that the longer the string the less will be the tension. Will the pressure on the nail be affected by the length of the string?

22. A string passing round a smooth peg is pulled at each extremity with a force equal to the strain on the peg; find the angle between the directions of the two portions of the string.

23. Four equal forces act on a particle. What are the conditions of equilibrium?

24. $ABDC$ is a parallelogram, and AB is bisected in E: prove that the resultant of the forces represented by AD, AC is double of the resultant of those represented by AE, AC.

25. The two systems of forces (P, Q, R) and (P, Q, R') at the same point are expressed by the "Parallelogram" and "Triangle of Forces" respectively. What is the relation between R and R'?

26. Four forces represented by AB, BC, CD, DE act on a point, and are balanced by the single force represented by AX. What is the position of X?

27. A point is taken within or without a quadrilateral, and lines are drawn from it to the angular points of the quadrilateral; prove that the resultant of the forces represented by these lines is represented by four times the line joining this point and the point of intersection of the lines joining the middle points of the opposite sides.

CHAPTER V.

ON THE RESOLUTION OF FORCES

41. IF we have a single force given, represented by AD, we can describe an infinite number of parallelograms, such as $ABDC$, having AD as diagonal.

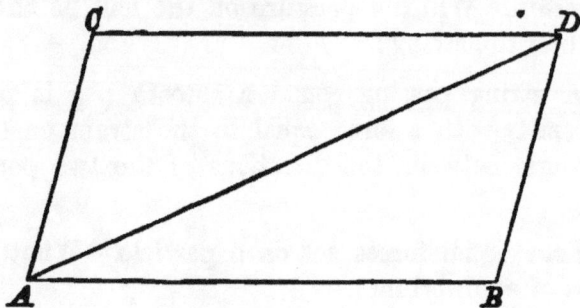

Hence we can obtain an infinite number of pairs of components, having as their resultant the force represented by AD.

42. If we have a single force given, represented by AD,

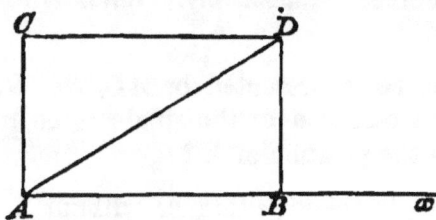

and we are required to find a pair of components, one of which shall act in a given direction Ax, we can describe an infinite number of parallelograms, such as $ABDC$, having AD as diagonal and one side AB lying along Ax.

Hence we can obtain an infinite number of pairs of components, of which one acts in the given direction.

43. But there is only one parallelogram having *AD* as diagonal, one side *AB* lying along *Ax*, and another side *AC* *making with* AB *a right angle at* A.

Hence we can obtain only one pair of components of a given force, such that one force acts in a given direction, and the other *perpendicular* to that direction. These are called the *resolved parts* of the given force in their own respective directions, and we give the following definition:

DEF. The resolved part of a given force in any direction is the force acting in this direction, which will, with another force acting in the perpendicular direction, have the given force for their resultant.

44. *To find the resolved part of the force represented by* AD *in the direction* Ax.

Drop *DB* perpendicular to *Ax*.

Complete the parallelogram *CABD*, having *CAB* a right angle.

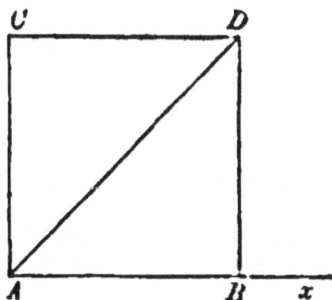

Then the forces represented by *AC, AB* act at right angles to each other, and have the force represented by *AD* for their resultant, and the force represented by *AB* acts along *Ax*; ∴ *AB* represents the resolved part of the force represented by *AD* in the direction *Ax*.

45. In Art. 46 we shall explain what the resolved part of a force in a given direction indicates *physically;* but first we must discuss the meaning of a phrase, which we shall have to employ in this explanation.

Suppose Ax and Ay to be two lines at right angles to each other, and a particle to travel along Az from A to z.

Draw zm, zn at right angles to Ax, Ay.

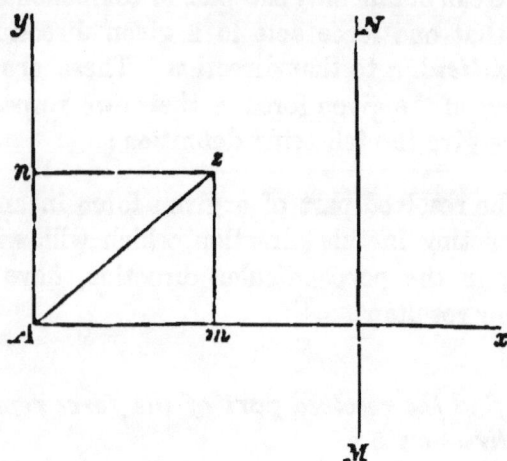

Then when the particle has reached z it has gone as far in the *direction* of Ax as if it had travelled along Ax from A to m. For if NM be a line perpendicular to Ax, the particle is, when at z, as much nearer NM, as it would be if it were at m. Similarly, when at z it has gone as far in the *direction* of Ay, as if it had travelled along Ay from A to n.

So that we may say that while travelling through Az, it has gone a distance equal to Am in the direction Ax, and a distance equal to An in the direction Ay.

46   Next, suppose the particle A to be acted on by a force R, represented in magnitude and direction by AD.

Drop perpendiculars DB, DC, on Ax and Ay.

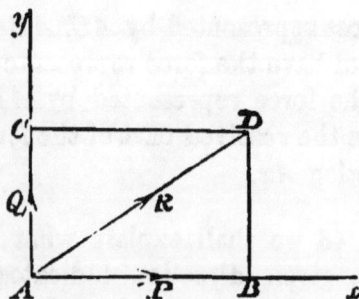

Then AB and AC will represent the resolved parts of R in the directions Ax, Ay; and these resolved parts we will call P and Q.

Now if the particle were acted on by Q alone, it would have no tendency to move in the direction Ax, and if by P alone, it would have no tendency to move in the direction Ay. But if it were acted on by R alone, it would have a tendency to move both in the direction Ax and also in the direction Ay. Therefore if it were acted on by P and Q together, it would have the same tendency; but of these Q, as we have seen, has no tendency to move it in the direction Ax. Hence P must have the same tendency to move it in the direction Ax that R has. In other words, the resolved part of a force R in a given direction Ax is the force, which by itself would give as much motion to a particle in the direction Ax, as R does in this direction.

We conclude, therefore, that the resolved part of a force in any direction represents the tendency which the given force has to move the particle acted on in that direction.

xlvii. Now suppose a force R to make with Ax an angle θ.

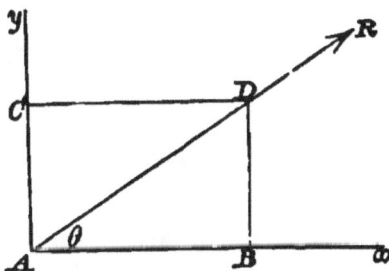

From any point D in R's line of action drop perpendiculars DB, DC, on Ax and Ay.

Then, if AD represent R in magnitude,

AB represents the resolved part of R in direction Ax,

AC .... Ay.

Now $AB = AD.\cos\theta$;

∴ resolved part of R in direction $Ax = R.\cos\theta$.

Hence we obtain the following rule :

To find the resolved part of a force in any given direction multiply the force by the cosine of the angle between the direction of the force and the given direction.

N.B.   The following is a case of frequent occurrence.   We have to find at the same time the resolved part of a force in *each* of two directions at right angles to one another.

Taking the notation of the last article,

resolved part of R in direction $Ax = R \cdot \cos \theta$,

resolved part of R in direction $Ay = R \cdot \cos \left(\dfrac{\pi}{2} - \theta \right) = R \cdot \sin \theta$.

xlviii.   *We may now proceed to find the resultant of any number of forces acting in one plane at a point.*

Through the point A draw two lines Ax, Ay at right angles to each other in the plane, in which the forces act.

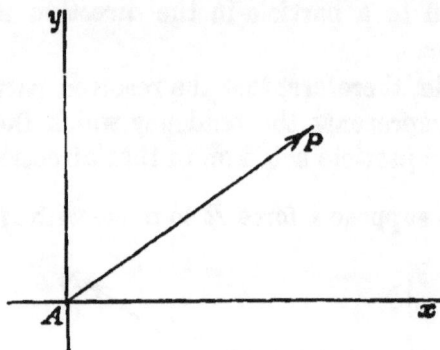

Let a be the angle which one of the forces, P, makes with Ax.

Then P is equivalent to $P \cdot \cos a$ acting in the direction of Ax, together with $P \cdot \sin a$ acting in the direction of Ay.

Similarly if P' be another of the forces, making an angle a' with Ax,

P' is equivalent to $P' \cdot \cos a'$ acting in the direction Ax,
 together with $P' \cdot \sin a'$ acting in the direction Ay.

Hence, for any number of forces P, P'...... making angles a, a'...... with Ax,
all the forces are equivalent to

$$P \cdot \cos a + P' \cdot \cos a' + \ldots\ldots \text{ in the direction } Ax,$$

together with $P \cdot \sin a + P' \cdot \sin a' + \ldots\ldots$ in the direction Ay.

For shortness' sake let $P \cdot \cos a + P' \cdot \cos a' + \ldots\ldots = X$,
 and $P \cdot \sin a + P' \cdot \sin a' + \ldots\ldots = Y$.

Also let R, the resultant of all the forces, make an angle θ with Ax.

$\therefore$ R is the resultant of X and Y,

which act at right angles to one another, and R makes an angle θ with X, and therefore

$R^2 = X^2 + Y^2$, which gives the magnitude of the resultant,

$\tan \theta = \dfrac{Y}{X}$, which gives the direction of the resultant.

xlix. *To find the conditions of equilibrium of a system of forces acting in one plane at a point.*

We have seen that the Resultant of any number of forces $P, P' \ldots \ldots$ may be determined in magnitude from the equation

$$R^2 = X^2 + Y^2,$$

where $X = P . \cos a + P' . \cos a' + \ldots \ldots$

and $Y = P . \sin a + P' . \sin a' + \ldots \ldots$

Now in order that $P, P' \ldots$ may be in equilibrium, their resultant must be zero (Art. 27):

that is, $R = 0,$

$$\therefore \ X^2 + Y^2 = 0.$$

But as the left-hand member of this equation consists of two terms, which, being squares, are essentially *positive*, their sum cannot be equal to 0, unless each be separately equal to 0;

that is, $X^2 = 0, \ \text{and} \ Y^2 = 0,$

and therefore $X = 0, \ \text{and} \ Y = 0,$

$$\therefore P . \cos a + P' . \cos a' + \&c. = 0,$$

and $P . \sin a + P' . \sin a' + \&c. = 0.$

These are the conditions of equilibrium, which may be expressed in words thus:

"The sums of the forces resolved *in any two directions in the plane of the forces at right angles to each other* must be severally *zero*."

EXAMPLES.—III.

1. Resolve a force of 12 lbs. into two forces, of which one is at right angles to it, and the other makes an angle of 30° with it.

2. Resolve a force of 10 lbs. into two forces, each making an angle of (1) 30°, (2) 60° with it.

3. What is the cosine of the angle at which the forces 9, 6, must act that their resultant may be equal to the greater force?

4. Shew how to resolve a given force into two others, (1) when they are of given magnitude, (2) when their directions are given.

5. Find the resultant of two forces of 13 lbs. each, represented by AB and AC, BC representing a force of 10 lbs.

6. If a force represented by AB be decomposed into two others represented by AC and AD, whereof $AC=AB$, shew that C and D lie on the circumferences of certain circles.

7. Shew that the sum of the squares of the resultants of two forces and of one of the forces and the other reversed is independent of the inclination of the forces.

8. Two pictures of equal weight are suspended symmetrically by cords passing over smooth pegs; the two portions of the cord in one case making an angle of 60°, and in the other an angle of 120° with each other : compare their tensions.

9. If the force 12 be decomposed into two others, 9 and 6, what are the cosines of their inclinations to the force 12?

10. A and B are fixed points on the circumference of a circle, P any other point on the circumference; shew that if two constant forces act along PA and PB, their resultant will pass through one point for all positions of P.

11. Two weights P and Q are joined together by a string, and laid on the circumference of a vertical semicircle, which is twice the length of the string. Find the position of equilibrium.

12. Show that if eight forces, acting on a particle, be represented in magnitude and direction by the straight lines drawn from the angular points of a quadrilateral to the middle points of the opposite sides, they will form a system in equilibrium.

13. AD is a quadrant of a circle, whose centre is O; B, C divide the arc AD into three equal parts; forces 8, 4, 4, 6 act along OA, OB, OC, OD respectively: determine their resultant.

14. Show that if four forces act at a point in the circumference of a circle, and be represented in magnitude and direction by the four straight lines drawn from that point to the angular points of a square inscribed in the circle, their resultant will be represented by four times the straight line drawn from the given point to the centre of the circle.

15. Through a point O, within a parallelogram $ABCD$, straight lines POQ, MON are drawn parallel to the sides, and meeting AB, BC, CD, DA in P, M, Q, N respectively: shew that if three forces, acting on a particle, be represented by PM, NQ, CA, they will form a system in equilibrium.

CHAPTER VI.

ON PARALLEL FORCES.

50. *To find the resultant of two forces, whose directions are parallel, acting on a rigid body.*

CASE I. When the forces act towards the same parts.

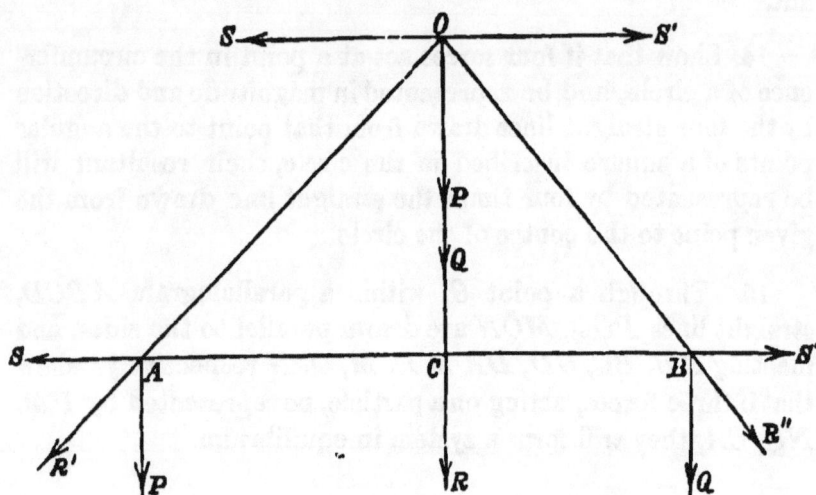

Let A, B be any two points in the lines of action of the two forces P, Q, acting in the parallel directions AP, BQ.

At A apply any force S in the direction BAS, and at B apply an equal force S' in the direction ABS': this will evidently not affect the combined action of the other forces.

Now S, P, acting at A, are equivalent to a single force R' acting in some direction AR'; and S', Q, acting at B, are equivalent to a single force R'', acting in some direction BR''. Let these two pairs of forces be replaced by R', R'', whose directions will meet in some point O, since SAR' and $S'BR'$ are together less than two right angles: and let the points of application of R', R'' be transferred to O.

Draw OCR parallel to AP and BQ, and SOS' parallel to AB.

Now let R', acting at O, be resolved into two components in directions OS and OC, which will clearly be S and P; and let R'', acting at O, be resolved into two components in directions OS' and OC, which will clearly be S' and Q.

Then S and S', being equal and opposite, will counteract each other, and may therefore be removed; and there will remain P and Q acting at O in the line OCR.

Hence if R be the resultant of P and Q, we know that it acts in a *direction* parallel to the directions of P and Q, and that its *magnitude* is $P + Q$.

Again, in the triangle ACO,

the sides are proportional to S, P, R', (Art. 38, Cor.);

and in the triangle BCO,

the sides are proportional to S'', Q, R'', (Art. 38, Cor.);

$$\therefore \frac{P}{S} = \frac{OC}{AC} \text{ and } \frac{S'}{Q} = \frac{BC}{OC};$$

$$\therefore \frac{P}{S} \times \frac{S'}{Q} = \frac{OC}{AC} \times \frac{BC}{OC};$$

$$\therefore \frac{P}{Q} = \frac{BC}{AC}.$$

Hence the resultant passes through a point C, which divides AB into segments inversely proportional to the forces.

CASE II. When the forces act towards opposite parts.

Let A and B be any two points in the lines of action of the two forces P, Q, which act in the parallel directions AP, BQ, and suppose Q to be greater than P.

At A apply any force S in the direction BAS, and at B apply a force $S'=S$ in the direction ABS'. Then S and P, acting at A, have a resultant R'; and S' and Q acting at B have a resultant R''.

Let the directions of R' and R'' meet in O. Draw SOS' parallel to AB.

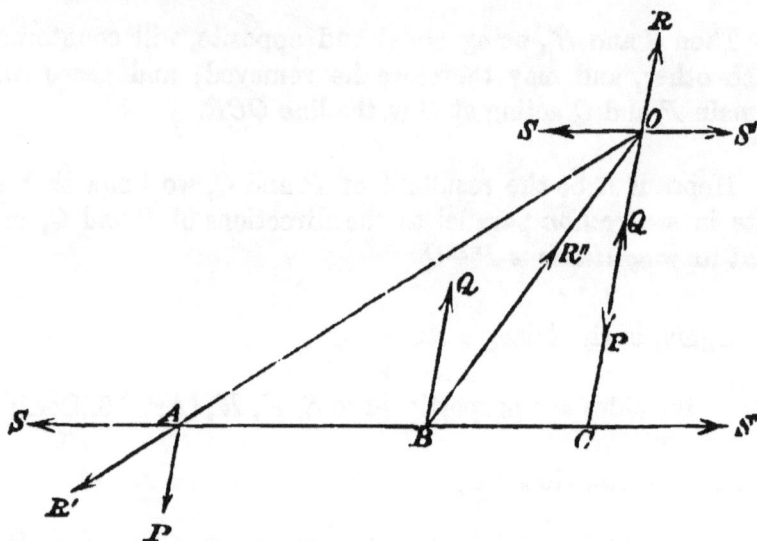

At O resolve the force R into two components, S, acting along OS, and P, acting along ROC, parallel to AP and BQ.

Also resolve R'' into two components, S' acting along OS', and Q acting along COR.

Then S and S', being equal and opposite, will counteract each other, and may be removed, and if R be the resultant of P and Q, there will remain a force acting along COR, such that

$$R=Q-P.$$

Also, ∵ the sides of △ ACO are parallel to P, R', S,
and the sides of △ BCO are parallel to Q, R'', S',

$$\therefore \text{(Art. 38. Cor.)}, \frac{P}{S} = \frac{OC}{AC}, \text{ and } \frac{S'}{Q} = \frac{BC}{OC}.$$

$$\therefore \frac{P}{S} \times \frac{S'}{Q} = \frac{OC}{AC} \times \frac{BC}{OC}, \text{ and } \therefore \frac{P}{Q} = \frac{BC}{AC}.$$

NOTE. We have supposed Q to be greater than P
We will now show that, this being the case, the diagram is
correctly drawn, that is, the directions of R' and R'' will
meet in a point O on the side of Q remote from P.

If two forces act at a point, the more one force is
increased, the smaller will be the angle between it and
the resultant.

Now, $S' = S$, and Q is greater than P, and ∴ R'', the
resultant of Q and S', makes with Q an angle less than
that, which R', the resultant of P and S, makes with P.

Hence ∠ OAC is less than ∠ OBC,
and ∴ ∠ OAC, OBA are together less than two right ∠'s:

∴ the directions of R' and R'' meet, when produced, on
the *upper* side of AB. But the part of the direction of
R'' on this side of AB lies also on the side of Q remote
from P.

∴ O must be on the side of Q remote from P.

EXAMPLES.—IV.

Suppose A and B to be two points in a rigid body, and P
and Q to be two parallel forces, acting in the same direction
through A and B, their resultant R passing through the
point C in the line AB, then solve the following questions:

(1) If $P = 20$ lbs., $Q = 30$ lbs., and $AC = 5$ inches, find BC.

(2) If $R = 28$ lbs., $P : Q = 3 : 4$, and $AB = 7$ inches, find AC.

(3) If $R = 14\frac{1}{2}$ lbs., $Q = 8$ oz., and $AC = 2$ inches, find AB.

If P and Q act in opposite directions, and C be a point in
AB produced, solve the following questions:

(4) $P = 3$ lbs., $Q = 5$ lbs., $AB = 12$ inches, find AC.

(5) $P = 10$ lbs., $AC = 18$ inches, $AB = 6$ inches, find Q.

51.   CENTRE OF PARALLEL FORCES.

The proposition, which we have just proved, enables us to find the resultant of any number of parallel forces, acting at different points of a rigid body.

For let P, Q, R,......be any number of parallel forces, acting at A, B, C,......points in a rigid body.

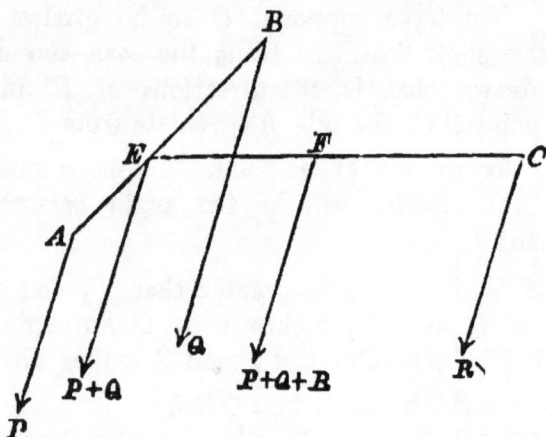

Then P and Q are equivalent to a single force $P+Q$, acting at a point E in the straight line AB, such that

$$P : Q = BE : AE.$$

Hence P, Q and R are equivalent to the parallel forces $P+Q$ and R, acting at the points E and C.

Then $P+Q$ and R are equivalent to a single force $P+Q+R$ acting at a point F in the straight line EC, such that

$$P+Q : R = CF : EF,$$

and thus any number of parallel forces may be reduced to a single resultant.

Now the position of the point, at which this resultant acts, does not depend on the direction, in which the component forces act, but only on their relative magnitude and their points of application. Hence if these component forces be turned about their points of application in any manner, still remaining parallel, the point, at which their resultant acts, will still be the same. For this reason that point is called *the centre of the parallel forces.*

52. *If three forces, acting upon a rigid body, balance each other, the lines, in which they act, must either be parallel or pass through a point.*

Fig. I.   Fig. II.

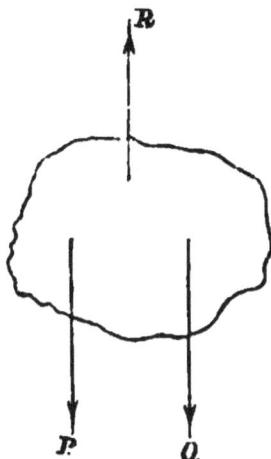
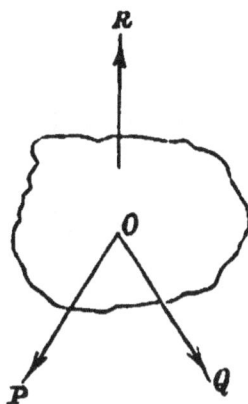

Let *P, Q, R* be the forces.

First, suppose *P* and *Q* to be parallel (as in fig. I.).

Then they will have some resultant, acting in a direction parallel to each of them. But this force, since it counteracts *R*, must be in direction exactly opposite to the direction in which *R* acts.

Consequently the line, in which *R* acts, must be *parallel* to the directions of *P* and *Q*.

Next, suppose the lines of direction of *P* and *Q* to meet in a point *O* (as in fig. II.).

Then the resultant of *P* and *Q* will pass through the point *O*.

But this force, since it counteracts *R*, must be in direction exactly opposite to the direction, in which *R* acts.

Consequently the line, in which *R* acts, must *pass through the point O*.

CHAPTER VII.

ON THE EQUILIBRIUM OF A BODY MOVEABLE ROUND A FIXED POINT.

53. If P and Q be two forces acting in one plane at points A and B of a rigid body, and O be a fixed point of the body, about which the body might turn freely, then, if under the action of P and Q the body be at rest, the resultant of P and Q must pass through O. For it is plain that any *single* force, acting on the body and not passing through O, will cause the body to turn about O.

Fig. I.   Fig. II.

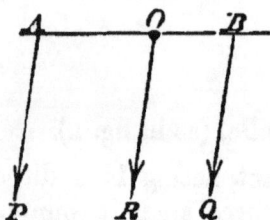 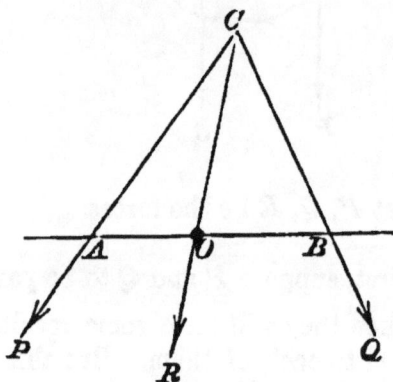

For the sake of simplicity let us suppose that the body in question is a rigid rod, without weight, moveable about a fixed point O.

Then, if P and Q be parallel, their resultant, R, will be parallel to both (as in fig. I.).

And if P and Q be not parallel, and C be the point, in which their lines of direction meet, COR will be the line of direction of the resultant of R (as in fig. II.).

54.   A rigid rod, capable of turning round a fixed point in the rod, is called a *Lever*.   The point, about which it can turn, is called the *Fulcrum*, and the parts, into which the rod is divided by the fulcrum, are called the *arms* of the lever. When the arms are in a straight line, the rod is called a *straight* lever: in all other cases it is called a *bent* lever.

55.   *If two forces, acting at right angles on a straight lever, produce equilibrium, the forces are inversely as their distances from the fulcrum.*

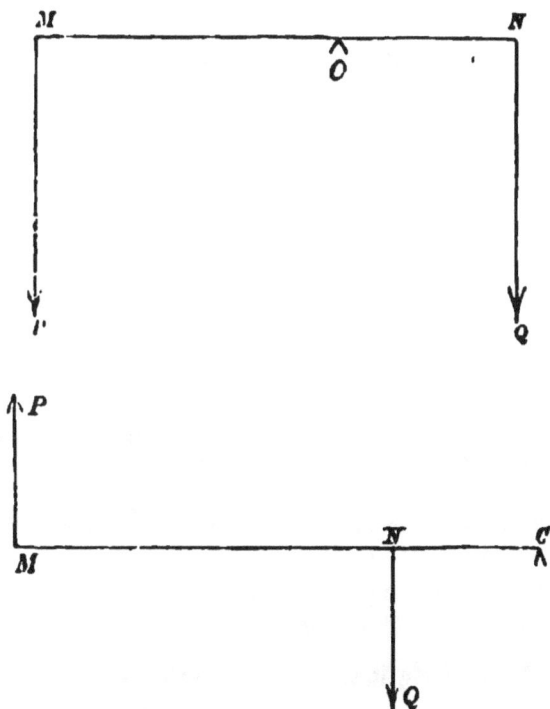

Let P and Q be the forces acting at the points M, N, and balancing each other round the fulcrum C.

Then the lever is kept at rest by P, Q, and the force exerted by the resistance of the fulcrum C.   Hence the resultant of P and Q must be equal and opposite to the force of this resistance, and must therefore pass through C.

Then, since C is the point, through which the resultant of P and Q passes, it follows from Art. 50, that

$$P : Q = CN : CM.$$

Also, if R be the pressure on the fulcrum,

in fig. 1, $R = P + Q$,

in fig. 2, $R = Q - P$.

56.   *If two parallel forces, acting on a lever, produce equilibrium, they are inversely as the perpendiculars drawn from the fulcrum to the directions in which the forces act.*

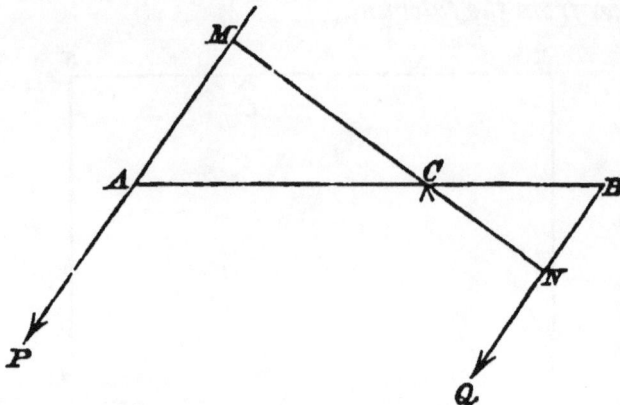

Let P and Q be two parallel forces balancing each other on the lever AB round the fulcrum C.

Then the lever is kept at rest by P, Q, and the force exerted by the resistance of the fulcrum C. Hence the resultant of P and Q must be equal and opposite to the force of this resistance, and must therefore pass through C.

Draw MCN at right angles to the directions of P and Q.

Then, since M and N are points in the lines of action of P and Q, and C the point in MN, through which the resultant of P and Q passes,

$$P : Q = CN : CM.$$

57.   *If two forces which are not parallel, acting on a lever, produce equilibrium, they are inversely as the perpendiculars drawn from the fulcrum to the directions in which the forces act.*

Let P and Q be two forces, not parallel, balancing each other on the lever ACB round the fulcrum C.

Produce the lines of direction of P and Q to meet in O.

Then the lever is kept at rest by P, Q, and the force exerted by the resistance of the fulcrum C. Hence the resultant of P and Q must be equal and opposite to the force of this resistance, and must therefore pass through C.

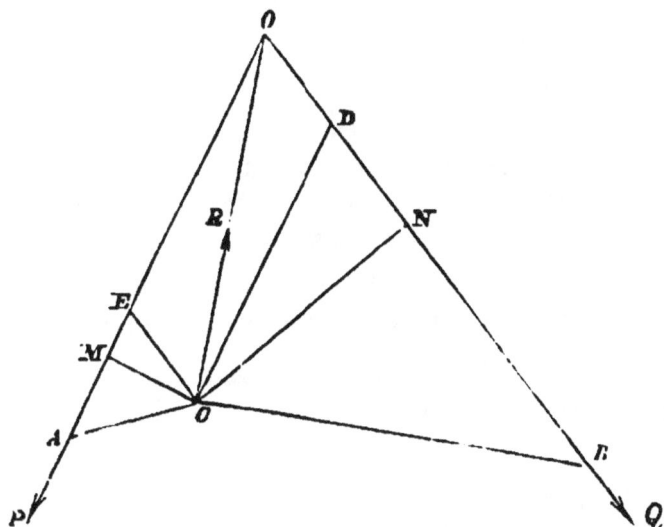

Let R be the resistance of the fulcrum; then the direction of R must pass through O. Art. 52.

Draw CD parallel to OP and CE parallel to OQ, and CM and CN at right angles to OP and OQ respectively.

Then the sides of the triangle CDO, being parallel to the directions of the three forces P, Q, R, may be taken to represent P, Q, R in magnitude. Art. 38.

Then
$$\frac{P}{Q} = \frac{CD}{OD} = \frac{CD}{CE}.$$

Now the triangles CME, CND are similar, since the right angles CME, CND are equal, and

angle CEM = angle DOE = angle CDN. (Eucl. I. 29.)

Hence
$$\frac{CD}{CE} = \frac{CN}{CM},$$

$$\therefore \frac{P}{Q} = \frac{CN}{CM}.$$

1. Weights of 5 lbs. and 7 lbs., hung at the extremities of a straight lever 6 feet long, balance each other: find the length of the arms.

2. A weight of 5 lbs., hung from one extremity of a straight lever, balances a weight of 15 lbs. hung from the other extremity: find the ratio of the arms.

3. When the pressure on the fulcrum is 99 lbs., and the arms of the lever are as 4 : 7, determine the weights.

4. The weight at the extremity of one arm of a straight lever is 12 lbs., the length of the arm is one foot, and the pressure on the fulcrum is 16 lbs.; what is the length of the arm at which the other weight acts?

5. If the weights on a lever are as 5 : 7, and the length of the lever is 36 inches, find the position of the fulcrum.

6. If two weights of 2 lbs. and 5 lbs. balance on a lever, one of whose arms is 5 inches longer than the other, compare the arms.

7. Two weights, P, Q, balance on a straight lever: if they be interchanged, determine the weight, which must be added to or subtracted from either to produce equilibrium.

8. Weights, proportional to 8 and 3, balance each other on a straight weightless Lever whose length is 2 feet 9 inches. Find the position of the Fulcrum: and state the ambiguity that would exist in the problem, if the equilibrium were produced by parallel forces instead of weights.

9. The weights on the extremities of a lever 8 feet long are as 1 to 3. Find the position of the fulcrum.

10. The length of a horizontal lever is 12 feet, and the balancing weights at the ends are 3 lbs. and 6 lbs. respectively. How far ought the fulcrum to be moved for equilibrium, if each weight be placed 2 feet from the ends of the lever?

11. A lever is 46 inches in length: the fulcrum is 8 inches from one end: what weight applied at this end will balance a weight of 22 lbs. at the other end?

12. At one end of a lever a fulcrum is placed, at the middle point of the lever hangs a weight of 15 lbs., and this is supported by an upward force of 10 lbs. acting 6 inches from the other end. Find the length of the lever.

13. A lever with a fulcrum at one end is 3 feet in length. A weight of 14 lbs. is suspended from the other end. At what distance from the fulcrum will an upward force of 25 lbs. preserve equilibrium?

14. A lever AB with a fulcrum at B is divided in the points C and D into three equal parts. From C and D weights of 12 lbs. are suspended. What force acting upwards at A will just support them?

15. AOB is a straight lever, and forces P and Q, acting at A and B, are in equilibrium about O, the fulcrum, which is 3 feet from A and 5 feet from B: solve the following questions:

(1) If $P=4$ lbs., $\angle BAP=90°$, $\angle QBO=60°$, find Q.

(2) If $P=3$ lbs., $\angle BAP=60°$, $\angle QBO=60°$, find Q.

(3) If $P=2$ lbs., $\angle BAP=45°$, $\angle QBO=30°$, find Q.

CHAPTER VIII.

ON THE CENTRE OF GRAVITY.

58. THE directions of the forces, which the Earth exerts on the different particles composing a body, are not, strictly speaking, parallel. But since the dimensions of any body, which we shall have to consider, are very small compared with its distance from the centre of the Earth, *we may consider these directions to be parallel.*

The resultant of this system of parallel forces is the weight of the body; and the point in the body, at which this resultant acts, is called *the Centre of Gravity of the body.*

Thus the Centre of Gravity is the centre of Parallel Forces, which act in a vertical direction.

We may suppose the whole weight of the body to be collected at the Centre of Gravity, and if it be in rigid connection with all the points in a body or a system of bodies, then the body or system would be in equilibrium in all positions, if the Centre of Gravity were supported.

Having thus explained the reasoning, on which we proceed to investigate the position of the Centre of Gravity of a body, we may give the following definition:

" The point at which the weight of a body or system may always be supposed to act is called the Centre of Gravity of the body or system."

59. *Every system of heavy particles has one and only one Centre of Gravity.*

Let A, B, C,...be any number of heavy particles,

P, Q. R,...the weights of A, B, C...

Suppose, first, that A and B are connected by a rigid rod without weight.

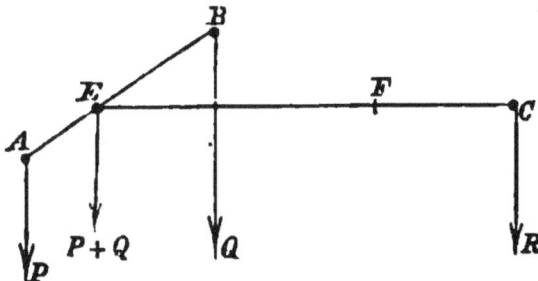

Now P and Q, being parallel forces acting in the same direction, are equivalent to a single resultant, the magnitude of which is $P + Q$, and which acts through a point E in the line AB, such that

$$P : Q = BE : AE.$$

E is then the centre of gravity of A and B, and the effect of P and Q will be the same as if A and B were collected into one particle, of weight $P + Q$, and placed at E.

Now suppose $P + Q$ to act at E: then we may find the centre of gravity of $P + Q$ acting at E and R acting at C, as before, by taking a point F in the line EC, such that

$$P + Q : R = CF : FE,$$

and we may suppose P, Q, R all collected at F: and so we may proceed for any number of particles.

Therefore every system of particles has a centre of gravity.

Also a system of particles can have *but one* centre of gravity.

For, if possible, let a system have two such points G and G', and let the system be turned about till the line joining G and G' is horizontal. Then we shall have the resultant of the system of parallel forces acting in a vertical line through G, and also in another vertical line through G'; which is impossible, since it cannot act in two different lines at the same time.

60. *To find the centre of gravity of a material straight line.*

Let AB be the given straight line.

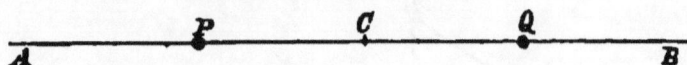

We may regard AB as a line of equal particles uniformly arranged. We may then divide the line into a series of pairs of equal particles, each pair being equidistant from C, the middle point of AB.

Let P and Q be such a pair of particles.

Then C will be the centre of gravity of P and Q.

Similarly each pair of the particles, of which AB is composed, will have C for its centre of gravity.

Therefore C will be the centre of gravity of the whole line AB.

61. *To find the centre of gravity of a parallelogram.*

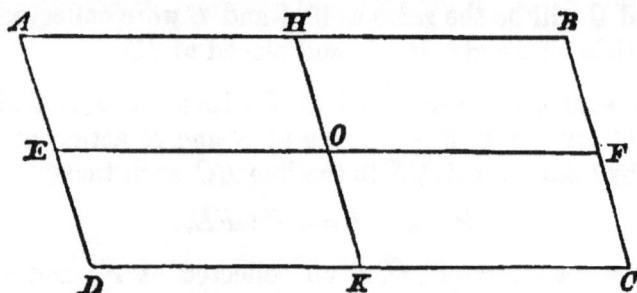

Let $ABCD$ be a parallelogram, regarded as a uniform lamina, *or thin sheet*, of matter. Draw EF parallel to AB and DC, bisecting AD, BC in the points E, F; and HK parallel to AD and BC, bisecting AB, DC in the points H, K.

We may suppose the parallelogram to be made up of a series of lines of particles, parallel to one of its sides, as BC.

Then it is plain that EF bisects each of these lines, and hence the centre of gravity of each of the lines composing the parallelogram is in EF.

Hence the centre of gravity of the whole parallelogram lies in EF.

Similarly it may be shewn to lie in *HK*.

Therefore *O*, the point of intersection of *EF*, *HK*, is the centre of gravity of the parallelogram.

62. *To find the centre of gravity of a plane triangle.*

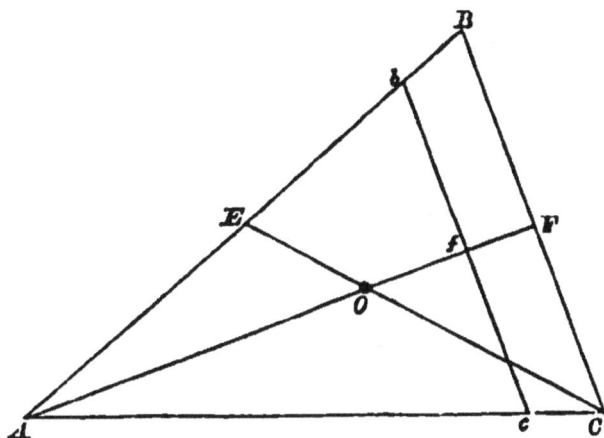

Let *ABC* be a plane triangular lamina of matter.

We may suppose this triangle to be made up of a series of lines of particles, running parallel to one of the sides, as *BC*.

Let *bc* be one of these lines.

Bisect *BC* in *F*, and join *AF*, cutting *bc* in *f*.

Now

$$Af : fb = AF : FB \quad \text{(by similar triangles } Afb, AFB\text{)}$$
$$= AF : FC \quad \text{(since } FB = FC\text{)}$$
$$= Af : fc \quad \text{(by similar triangles } AFC, Afc\text{)};$$
$$\therefore fb = fc.$$

Similarly it may be shewn that *AF* will bisect each of the lines parallel to *BC*; and hence the centre of gravity of each of the lines composing the triangle is in *AF*, and therefore the centre of gravity of the triangle is in *AF*.

Now bisect *AB* in *E* and join *CE*.

Then the centre of gravity of the triangle will be in *CE*.

Therefore the point *O*, in which *AF* and *CE* cut each other, will be the centre of gravity of the triangle.

63.   *To shew that if a line be drawn from any angle to the middle point of the opposite side, the centre of gravity of the triangle lies in this line at a distance from the angular point equal to two-thirds of the line.*

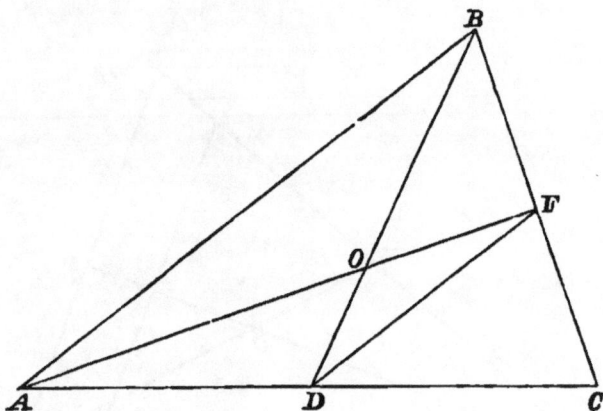

Draw BD and AF to the middle points of AC and BC.

Then O, the intersection of AF, BD, is the centre of gravity of the triangle.

We have now to shew that $BO = 2OD$.

Join FD.

Then, since FD bisects BC and AC, it must be parallel to AB.   (Eucl. vi. 2.)

And since ABC, DFC are similar triangles,

$$AB : DF = AC : DC$$

$$= 2 : 1.$$

Again, since AOB, FOD are similar triangles,

$$BO : OD = AB : DF$$

$$= 2 : 1;$$

$$\therefore BO = 2OD.$$

And hence BO is two-thirds of BD.

64. *The centre of gravity of a triangle coincides in position with the centre of gravity of three equal particles placed at the angular points.*

Let three particles, each of weight P, be placed at A, B, C.

Take D the middle point of AC.

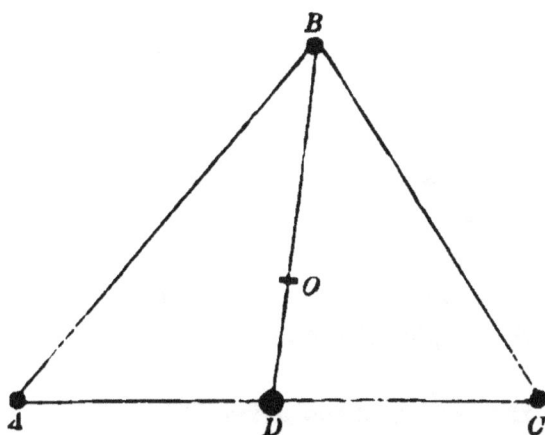

Then D will be the centre of gravity of the particles acting at A and C, and we may suppose both to act at D.

Then we have $2P$ acting at D, and P at B, and the centre of gravity of these weights will be found by joining BD and taking in it a point O, such that

$$BO : OD = 2P : P$$

$$= 2 : 1.$$

i.e. O is the centre of gravity of the triangle.

65.   *If a body be suspended from a point, about which it can swing freely, it will rest with its centre of gravity in the vertical line, which passes through the point of suspension.*

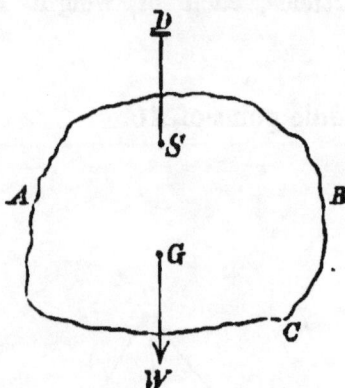

Let *ABC* be the body, *G* its centre of gravity, *S* the point in the body, by which it is suspended by a string fastened to the fixed point *D*.

Then there are two forces acting on the body:

      (1)   the weight of the body acting in the *vertical* line *GW*,

      (2)   the tension of the string *DS*.

When the body is at rest, these two forces must act in the same straight line;

    ∴ *DS* is a vertical line;

    ∴ *G* is in the vertical line passing through *S* and *D*.

66.   *The position of the centre of gravity of a body may be sometimes determined by experiment in the following manner:*

Let the body be suspended from any point in its surface, and let the line, which is vertical and passes through the point of suspension, be marked.

Then let the body be suspended from another point in its surface, and let the line, which is vertical and passes through the point of suspension, be marked.

The point of intersection of the two lines in the body is the centre of gravity of the body.

67. *A body, placed on a horizontal plane, will stand or fall over, according as the vertical line drawn through the centre of gravity of the body falls within or without the base.*

Fig. I.

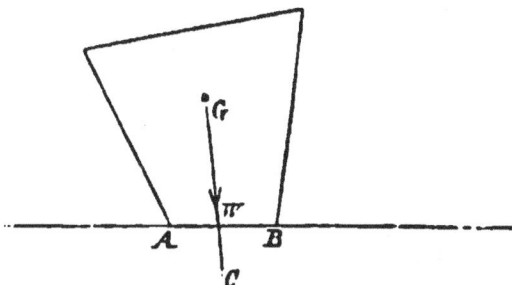

Suppose the vertical line GC, passing through the centre of gravity G to fall within the base, as in fig. I. Then we may suppose the weight of the body (W) to be concentrated at G. There will then be a vertical pressure of W downwards acting in the line GC, which will be counteracted by an equal and opposite pressure of the plane, on which the body is placed, acting upwards in the direction CG, and so equilibrium will be produced, and the body will stand.

Fig. II.

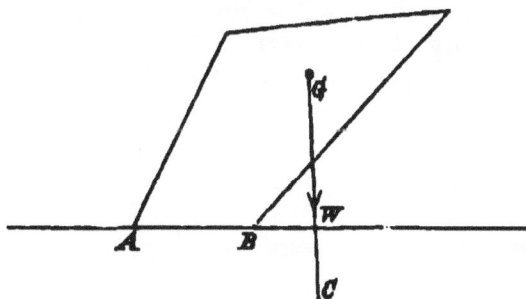

But suppose, as in fig. II., that the line GC falls without the base: then there is no pressure equal and opposite to W, and the body will be turned round B, the nearest point of contact in the base to the vertical line GC, and will fall.

N.B. By the *base* is here meant the figure bounded by a string drawn tightly round the parts of the body in contact with the horizontal plane.

Thus the base on which a chair stands is the quadrilateral of which the feet are the four corners.

68. The following illustration may enable the learner to grasp more completely the fact enunciated in the preceding Article.

Cut a piece of cardboard of uniform thickness in the form of a square.

In the corners *A, B, C* insert three pins of equal length and place them in a vertical position with their points *M, N, O* resting on a horizontal table. The centre of gravity of the cardboard is *G* the middle point of *AC*, and this point is vertically above *H* the middle point of *MO*.

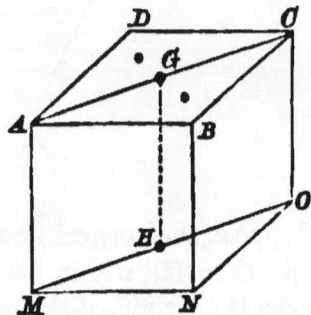

If now we put a small weight on the triangle *ABC*, the centre of gravity of the cardboard and weight will be vertically above some point *within* the triangle *MON*, and the system will stand.

But if we put the small weight on the triangle *ACD*, the centre of gravity of the cardboard and weight will be vertically above some point *outside* the triangle *MON*, and the system will fall.

69. *On stable and unstable equilibrium.*

(1) If a body under the action of any force be in a position of equilibrium, and a *very small* displacement be given to the body, if it then tend to return to the original position of equilibrium, that position is called one of *stable equilibrium*.

(2) If the body tend to move further from its original position, that position is called one of *unstable equilibrium*.

(3) If it remain in the new position which the displacement has given it, the position is said to be *neutral*.

Examples.

(1) A weight suspended by a string is an instance of stable equilibrium.

(2) A stick balanced on the finger is an instance of un-stable equilibrium.

(3) A sphere resting on a horizontal table is an instance of neutral equilibrium.

70. *Having given the centre of gravity of a body and also the centre of gravity of a part of the body, to find the centre of gravity of the remainder.*

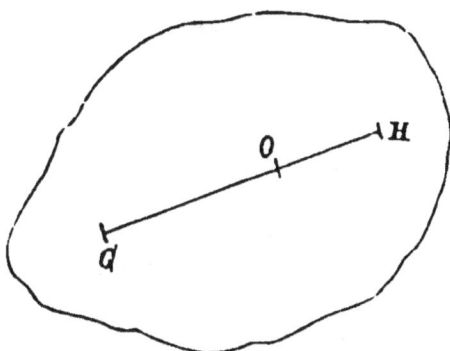

Let G and H be the centres of gravity of the two parts of the body, w and x the weights of the parts respectively.

Then, if O be the centre of gravity of the whole body,

$$GO : OH = x : w;$$

$$\therefore w \times GO = x \times OH;$$

$$\therefore OH = \frac{w \times GO}{x}.$$

Hence if G and O be given we can find H, by producing GO so far that the part produced $= \dfrac{w \times GO}{x}$.

71.   *To find the centre of gravity of a number of particles lying in a straight line.*

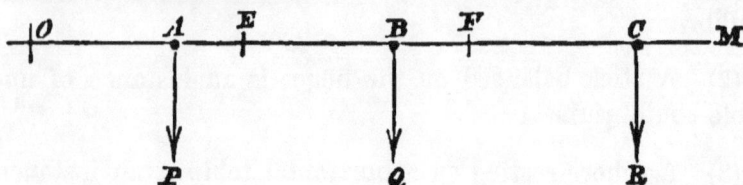

Let A, B, C... be the several particles lying in the straight line OM:

P, Q, R...their weights.

Let O be a fixed point in the line.

Then P at A and Q at B have a centre of gravity at E, such that

$$P . AE = Q . BE.$$

Then $P . (OE - OA) = Q . (OB - OE)$,

or, $P . OE + Q . OE = P . OA + Q . OB$,

$$\therefore OE = \frac{P . OA + Q . OB}{P + Q}.$$

Next, the two weights, $P + Q$ at E and R at C, have a centre of gravity at F, such that

$$OF = \frac{(P + Q) . OE + R . OC}{(P + Q) + R};$$

$$\therefore OF = \frac{P . OA + Q . OB + R . OC}{P + Q + R}.$$

And so on for any number of particles.

72. *To find the centre of gravity of the perimeter of a triangle—regarding the sides as material lines of uniform thickness.*

Let *D, E, F* be the middle points of the sides of the proposed triangle *A BC*.

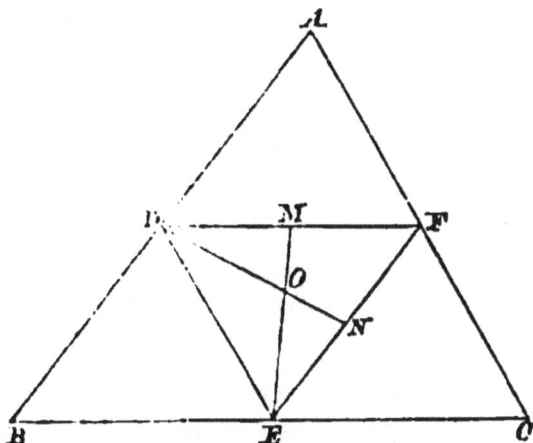

Then the centre of gravity of the perimeter *ABC* will be in the same position as the centre of gravity of three particles placed at *D, E, F,* whose weights are proportional to *AB, BC, CA* respectively.

Draw *EM* bisecting the angle *DEF*, and *DN* bisecting *EDF*.

Now $DM : MF = DE : EF$, by Euclid, vi. 3,
$$= AC : AB, \text{ by sim}^r. \text{ triangles } ABC, EFD$$

Hence *M* is the centre of gravity of the two sides *AB, AC*; and therefore the centre of gravity of the whole perimeter lies in *EM*.

Similarly it lies in *DN*.

Therefore *O* the point of intersection of *EM, DN* is the centre of gravity required, and this, by Euclid, iv. 4, is the centre of the circle inscribed in the triangle *DEF*.

Examples. - VI.

1. A uniform rod, of 3 feet in length and 8 lbs. weight, has a pound weight placed at one end; find the centre of gravity of the whole system.

2. A ball of weight 2 lbs. lies in the middle between two balls, a foot apart, of weights 4 lbs. and 1 lb.; find the centre of gravity of the three.

3. Four equal particles are placed in a straight line, the distance between the first and second being one inch, between the second and third two inches, and between the third and fourth three inches; find their centre of gravity.

4. Find the centre of gravity of weights whose measures are 16, 8, 4, 2, 1, placed at distances of 1 ft. from each other in a straight line.

5. Show that if the centre of gravity of three heavy particles, placed at the angular points of a triangle, coincide with the centre of gravity of the triangle, the particles must be of equal weight.

6. Show that if a number of triangles be described upon the same base and between the same parallels, their centres of gravity lie on a straight line.

7. If the angular points of one triangle lie at the middle points of the sides of another, show that the triangles will have the same centre of gravity.

8. If two triangles are upon the same base, show that the line joining their centres of gravity is parallel to the line joining their vertices.

9. Two equal particles are placed on two opposite sides of a parallelogram; show that their centre of gravity will remain in the same position, if they move along the sides so as to be always equidistant from opposite angles.

10. Having given the positions of three particles A, B, C, and the position of the centres of gravity of A, B and A, C, find the position of the centre of gravity of B, C.

11. An equilateral triangle is inscribed in a circle; show that its centre of gravity will be the centre of the circle.

12. If the centre of gravity of a triangle inscribed in a circle coincide with the centre of the circle, show that the triangle is equilateral.

13. Why does a man, when he is carrying a weight with one arm, extend the other?

14. A uniform flat rod whose length is 14 inches and weight 3 lbs. rests on a horizontal table; if a weight of 4 lbs. be placed on one end of the rod, find the greatest distance, which the other end may be made to project beyond the table, without the rod falling off.

15. Find the centre of gravity of four equal particles A, B, C, D, when the straight line AB bisects the straight line CD.

16. A triangle suspended from one of its angles has its base horizontal; show that the triangle is isosceles.

17. A square and a triangle have the same base; find the altitude of the triangle, if the centre of gravity of the two lie in that base.

18. A right-angled triangle is suspended freely against a wall from its right angle; and again from one of the other angles. If in these two cases the positions of the hypotenuse be at right angles to each other, compare the length of the sides of the triangle.

19. If ABC be an equilateral triangle, and weights 3 lbs., 4 lbs., 5 lbs. be placed at its corners, find their centre of gravity.

20. Find the centre of gravity of three weights at the angular points of an equilateral triangle, two of the weights being each double of the third.

21. If an equilateral parallelogram be suspended from an angular point, one diagonal will be horizontal.

22. If a parallelogram be divided into four triangles by its diagonals, and the centres of gravity of the four triangles be joined, the joining lines will form another parallelogram.

23. If a triangle ABC, right-angled at C, and having the side opposite A double that opposite B, be suspended successively by A and C from a peg in a vertical wall, find the angle between the two positions of AC.

24. A right-angled triangle, whose acute angles are to each other as 1 : 5, is suspended from the right angle; determine the inclination of the hypotenuse to the vertical.

25. An isosceles triangle is suspended successively from the angular points of its base; shew that the two positions of the base will be at right angles, if the base of the triangle be two-thirds of its altitude.

26. AB is a straight line, C a point in it such that $AC = 2CB$. Weights of 1 lb., 2 lbs., 3 lbs. are placed at A, B and C respectively. Find their centre of gravity.

27. $ABCD$ is a straight line divided into three equal parts at the points B and C, and equal heavy particles are placed at the points A, B, D. Find their centre of gravity.

28. $ABCD$ is a parallelogram having the angle ABC = 60°, and the base BC six inches in length, what length must the side AB not exceed, if the figure is to stand upon BC?

29. O is the centre of a circle, AOB a diameter; C the middle point of the arc AB; D, E points on the arcs AC, CB respectively, at distances from AOB each equal to half the radius AO. Equal heavy particles are placed at O, A, B, C, D, E. Find their centre of gravity.

30. If a body, whose centre of gravity is G, be divided into two parts, and if G_1, G_2 be the centres of gravity of the two parts, shew that $G_1 G G_2$ is a straight line.

31. AB the base of a square $ABCD$ is divided in E and the triangle CBE removed; shew that the remainder will stand or fall according as BE bears to EA a less or greater ratio than $\sqrt{3} : 1$.

32. Find the centre of gravity of a trapezium, of such form, that the line joining one pair of its opposite angles divides it into two equal triangles.

33. From a given square shew how to cut a triangle having one side of the square for its base, so that the centre of gravity of the remaining portion may be at the vertex of the triangle.

34. If a triangle have a side upon which it will not stand, that side is the shortest, and is not half the length of the longest side.

35. If a quarter of a trianglo be cut off by a line drawn parallel to one of its sides, find the centre of gravity of the remainder

36. In a right-angled triangle, of which the greatest side is 6 inches long, what will be the distance of its centre of gravity from the right angle?

37. *ABCD* is a square surmounted by an isosceles triangle on one side *BC*, which forms the base of the triangle. Determine the greatest height of this triangle, so that the figure (which is cut out of one piece of board) can be placed on its side *DC* without tumbling over.

38. Find the position of the centre of gravity of a figuro formed by an equilateral triangle and a square, the base of tho trianglo coinciding with one of the sides of the square.

39. Find the contre of gravity of the quadrilateral formed by drawing a straight line parallel to the base of an isosceloi triangle and bisecting tho sides.

40. A square is divided into 9 smaller squares. At the centre of each of the latter a weight is placed of tho magnitude indicated in the margin. Find their centre of gravity. Suppose the weight 9 to be removed; where will be the centre of gravity of the rest?

4	9	2
3	5	7
8	1	6

41. Tho diagonal of a square *ACEG* is divided in *K* so that *GK* = 2*KC*. Through *K* are drawn two lines parallel to tho sides of the square, meeting the sides *AC* in *B*, *CE* in *D*, *EG* in *F*, *GA* in *H*. At all the angles of the figure are placed weights, 1 lb. at *A*, 2 lbs. at *B*, 3 lbs. at *C*, 4 lbs. at *D*, 5 lbs. at *E*, 6 lbs. at *F*, 7 lbs. at *G*, 8 lbs. at *H*, 9 lbs. at *K*. Find their centre of gravity.

42. Construct a figure as in Ex. 41, except that no weights are to be placed at the angles: remove the portion *HKFG*: suppose matter distributed uniformly over tho remainder: find the centre of gravity.

43. A heavy bar 14 feet long is bent into a right anglo, the legs of which are 8 feet and 6 feet long respectively; provo that the distance of the centre of gravity of the bar so bent from the point in it, which was its centre of gravity when it was straight, is $\frac{9\sqrt{2}}{7}$ foot.

44. A wire is bent into the form of a hexagon with one side omitted. Find its centre of gravity.

45. A uniform triangular board of given weight rests in a horizontal position on three props; find the portion of the weight sustained by each.

46. A heavy circular disc is suspended from its centre by a string and rests horizontally. Weights of 3, 3 and 5 lbs. are suspended by strings, attached to the middle point of the disc and passing over its edges. Shew how to arrange them, so that the disc may remain horizontal.

47. Assuming the position of the centre of gravity of a triangle, find that of any quadrilateral cut off by a straight line parallel to one side.

48. If a parallelogram suspended from an angular point have one diagonal horizontal, prove that the four sides are equal.

49. Find the centre of gravity of weights, w, $2w$, $3w$ placed at the angular points of a triangle, and determine the ratios, in which the lines, drawn from the angular points through the centre of gravity to the opposite sides, are divided at that point.

50. An isosceles triangular lamina weighs 6 lbs. What weight, placed at the vertex, will make the triangle balance about the middle point of the perpendicular from the vertex on the base?

51. ABC is a right-angled triangle without weight, A the right angle: place such weights at B and C, that, when the triangle is suspended from A, it may rest with the side BC horizontal.

52. A cubical block of stone of uniform density rests on a rough horizontal plane. It is tilted up, by having a wedge driven under it with its edge parallel to an edge of the cube. Shew that the block will not be overturned, unless the angle of the wedge be greater than 45°.

53. Find the centre of gravity of three squares described on the sides of an isosceles right-angled triangle.

CHAPTER IX.

OF MOMENTS.

73. WE have hitherto treated chiefly of the tendency of forces to produce motion of a particle or body *away from* a fixed point, that is, to produce what is termed *displacement by translation.*

We shall now have to consider, in extension of the principles laid down in Chap. V., the tendency of forces to produce motion *round* a fixed point, that is, to produce what is termed *displacement by rotation.*

74. The MOMENT of a force about a point is the name given to the power, which that force has to turn any body round that point.

In estimating the power of a force to turn a body round a *fixed* point, we shall have to take into account

(1) The magnitude of the force.

(2) The perpendicular distance of the line of action of the force from the *fixed* point.

We must first explain what we mean by the *measure* of each of these elements, and then we shall be able to explain how the moment of a force is *measured.*

We measure the *magnitude* of a force by the *number* expressing how many times the force contains a certain force, selected as the unit of force.

Thus, when we speak of a force F, we mean a force which contains the unit of force F times.

We measure the perpendicular *distance* of a line from a point by the *number* expressing how many times the perpendicular distance contains a certain line, selected as the unit of distance.

Thus, when we speak of a perpendicular distance D, we mean a distance which contains the unit of distance D times.

75.   Now suppose A, B, C, D... to be any number of rods *of equal length*, fixed in a circular wheel moveable about its centre.

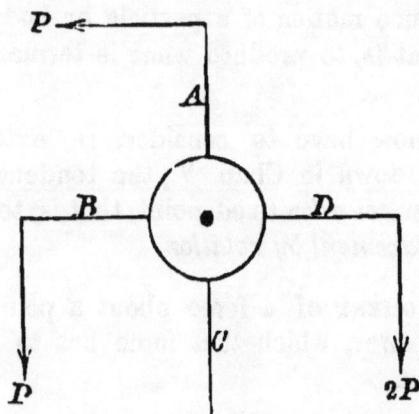

Then it is evident that the power of a force P to turn the wheel will be the same, when applied perpendicularly at the extremities of any one of the rods.

Also, if equal forces P and P be applied perpendicularly at the extremities of the rods A and B, so as to turn the wheel from *right to left*, it is plain that the rotatory power of these forces will be just counteracted by a force $2P$, applied perpendicularly at the extremity of the rod D, acting in a contrary direction, that is, tending to turn the wheel from *left to right*.

Hence we infer that a force $2P$, acting at the extremity of D, will have *twice* the rotatory effect of a force P, applied in the same direction at the same point.

And thus we conclude *generally* that, if two forces F_1 and F_2 be applied at the same point of a rod in the same direction,

$$\text{moment of } F_1 : \text{moment of } F_2 = F_1 : F_2,$$

that is, the moment of any force acting at a constant distance varies as that force.

76.   *If F be the measure of a force acting on a body, and D the measure of the perpendicular distance of the line*

of action of the force from a fixed point in the body, the measure of the moment of the force about the point is equal to F . D.

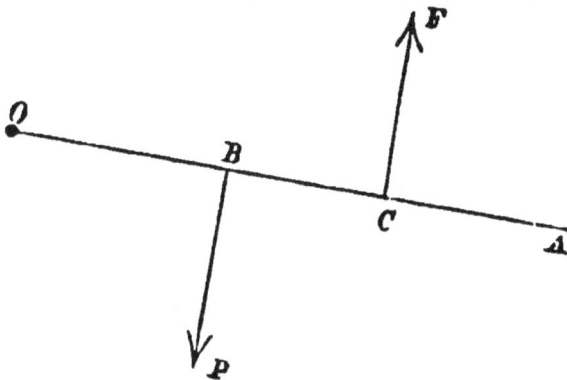

Let O be the fixed point, OA the straight line through O, perpendicular to the direction of F, and meeting it in C, so that the measure of OC is D.

Let B be a point in OA, whose distance from O is the same, whatever F and D may be.

Let P be the force, which, when applied at B perpendicularly to OA, will counteract the effect of F.

Hence, moment of F about O = moment of P about O.

Also, the resultant of the parallel forces F and P must pass through O,

$$\therefore P . OB = F . OC, \qquad \text{(Art. 50)}$$

and $\therefore P$ varies as $F . OC$, since OB is constant.

But moment of P varies as P, ... (Art 75);

$\therefore$ moment of F varies as $F . OC$,

that is, moment of F varies as $F . D$.

Now, if we take, as our unit of moment, the moment of a unit of force about a point at a unit of distance from the direction of the force,

moment of F : unit of moment = $F . D : 1$;

$\therefore$ moment of F is $F . D$ times the unit of moment;

$\therefore$ the measure of the moment of $F = F . D$.

Cor. Hence the moment of a force about a point in its own line of action is *zero*

77. We have shewn that forces can be represented geometrically by straight lines, and we can now shew that moments can be represented geometrically by *areas*.

Let us suppose that a force P, represented in magnitude and direction by the line AB, acts at A, a point in a straight rod AO, moveable about O.

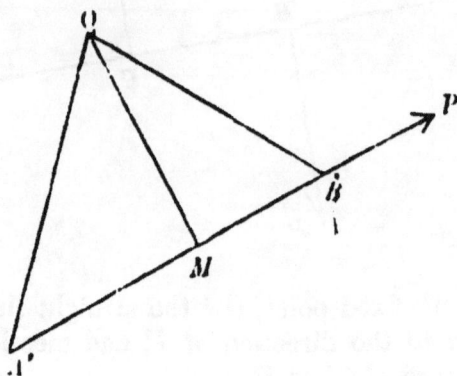

Draw OM at right angles to the line of action of P, and join OB.

Then, since the force P contains P units of force, the line AB must contain P units of length.　(Art. 16.)

∴ the measure of AB is P.

Let the measure of OM be D.

Then the measure of the area of the triangle AOB is $\dfrac{P \cdot D}{2}$;

∴ measure of *twice* the area of triangle AOB is $P \cdot D$.

Also, measure of moment of P about O is $P \cdot D$.　(Art. 76.)

Hence twice the area of the triangle AOB will represent the moment of P about O.

We conclude, then, *that the moment of a force about a point may be represented geometrically by twice the area of the triangle, whose vertex is the point, and whose base is a line representing the force in magnitude and direction.*

78. Let us now take the case of two forces *P* and *Q*, represented by the lines *AB*, *AC*, acting at *A* on the rod *AO* in such a way, that the perpendicular drawn from the fixed point *O* to *P*'s line of action falls *on one side* of *AO*, and the perpendicular drawn from *O* to *Q*'s line of action falls *on the other side* of *AO*

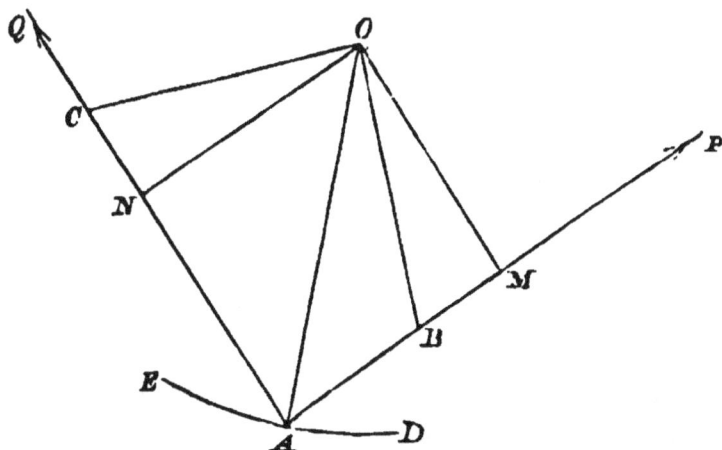

The moment of *P* about *O* will then be represented by twice the triangle *AOB*, and the moment of *Q* about *O* will be represented by twice the triangle *AOC*.

Now the force *P* tends to cause the point *A* to move along the circular arc *AD*, and the force *Q* tends to cause the point *A* to move along the circular arc *AE*.

Thus the forces tend to turn the rod *AO* in contrary directions, and this difference we can express by the terms *positive* and *negative*. These terms are only relative, and may be applied, at discretion, to express causes or effects that are directly opposed to each other, but for convenience sake we make the following statement:

The moment of a force may be considered negative or positive, according as the force tends to turn the body in the same direction as that, in which hands of a watch revolve, or the contrary.

79. *The algebraic sum of the moments of two parallel forces about any point in their plane is equal to the moment of their resultant about that point.*

CASE I. When the forces act in *the same* direction.

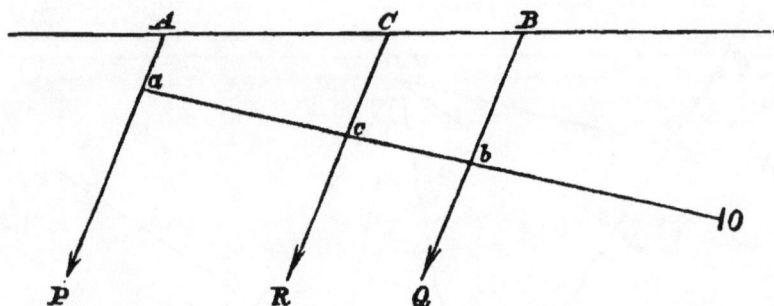

Let A, B be two points in the line of action of P, Q; and let C be the point in the line AB, through which R, the resultant of P and Q, passes.

Take any point O in the plane of the forces.

Draw *Obca* at right angles to the directions of the forces.

Then we may suppose P, Q, R to act at a, b, c (Art. 17), and therefore, by Art. 50,

$$P : Q = bc : ac;$$

$$\therefore P \cdot ac = Q \cdot bc \ldots\ldots\ldots\ldots (1).$$

Then, algebraic sum of moments of P and Q round O

$$= P \cdot Oa + Q \cdot Ob \cdot$$

$$= P \cdot (Oc + ac) + Q \cdot (Oc - bc)$$

$$= P \cdot Oc + P \cdot ac + Q \cdot Oc - Q \cdot bc$$

$$= P \cdot Oc + Q \cdot Oc, \qquad \text{from (1)}$$

$$= (P + Q) \cdot Oc$$

$$= R \cdot Oc$$

$$= \text{moment of } R \text{ round } O.$$

CASE II. When the forces act in *opposite* directions.

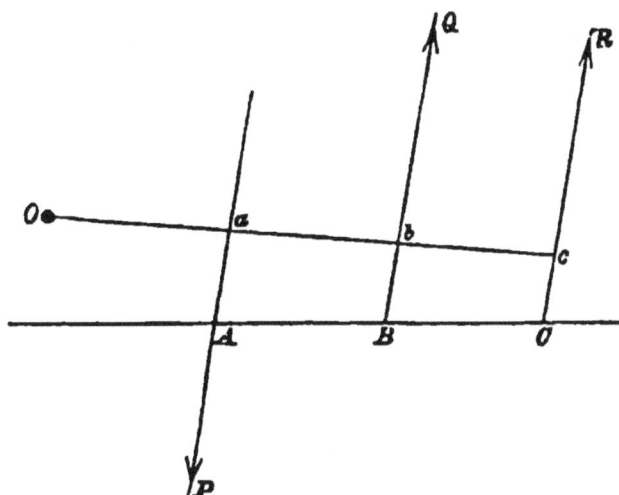

Let P and Q be the forces, of which Q is the greater.

Let A, B be two points in the lines of action of P, Q, and let C be the point in the line AB produced, through which R, the resultant of P and Q, passes.

Take any point O in the plane of the forces, and draw $Oabc$ at right angles to the directions of the forces.

Then we may suppose P, Q, R to act at a, b, c (Art. 17), and therefore, by Art. 50,

$$P : Q = bc : ac \cdot$$
$$\therefore P \cdot ac = Q \cdot bc \dots \dots \dots \dots (1).$$

Then, the moment of P round O being *negative*, Art. 78, algebraic sum of moments of P and Q round O

$$= Q \cdot Ob - P \cdot Oa$$
$$= Q \cdot (Oc - bc) - P \cdot (Oc - ac)$$
$$= Q \cdot Oc - Q \cdot bc - P \cdot Oc + P \cdot ac$$
$$= Q \cdot Oc - P \cdot Oc, \qquad \text{from (1)}$$
$$= (Q - P) \cdot Oc$$
$$= R \cdot Oc$$
$$= \text{moment of } R \text{ round } O.$$

80. *The algebraic sum of the moments of two forces, meeting in a point and acting in one plane, about any point in the plane, is equal to the moment of their resultant about that point.*

CASE I. When the point is *within* the angle between the forces.

Let AB, AC represent the two forces P, Q.

Complete the parallelogram $ABDC$.

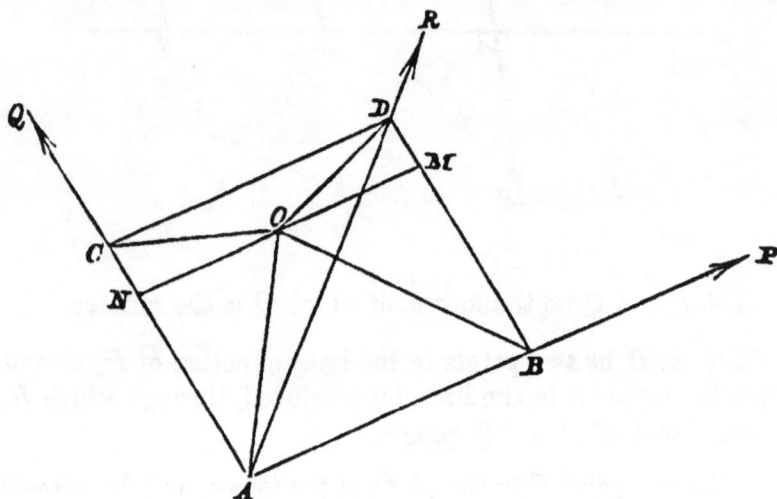

Draw MON parallel to AB and CD.

Now, as the figure is drawn, the moments of P and R about O are *positive*, and the moment of Q about O is *negative*, and we have to shew that

2 triangle AOB – 2 triangle AOC = 2 triangle AOD.

Now

parallelogram BN = parallelogram BC – parallelogram MC;

∴ 2 triangle AOB = 2 triangle ADC – 2 triangle DOC

$$= 2 \, (\text{triangle } ADC - \text{triangle } DOC)$$

$$= 2 \, (\text{triangle } AOC + \text{triangle } AOD)$$

$$= 2 \text{ triangle } AOC + 2 \text{ triangle } AOD;$$

∴ 2 triangle AOB – 2 triangle AOC = 2 triangle AOD, which proves the proposition.

CASE II. When the point is *without* the angle between the forces.

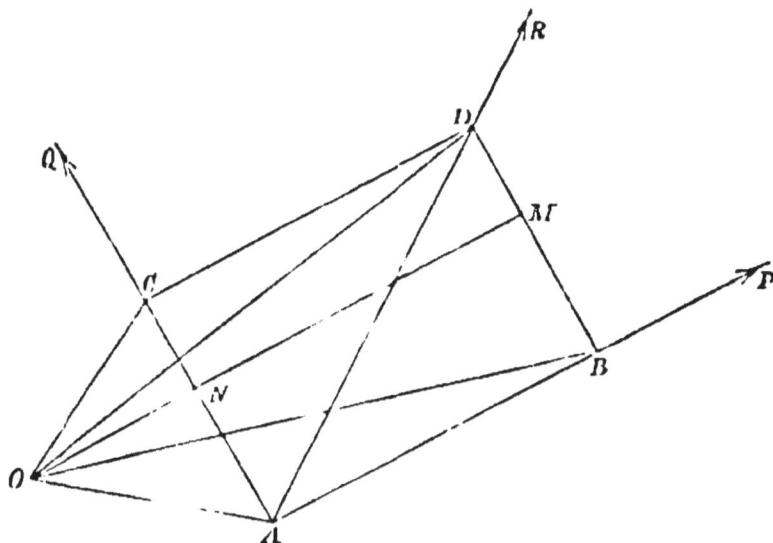

Draw *ONM* parallel to *OD* and *AB*.

As the figure is drawn, the moments of *P*, *Q*, *R*, about *O* are all *positive*, and we have to shew that

2 triangle *AOB* + 2 triangle *AOC* = 2 triangle *AOD*.

Now 2 triangle *AOD*

= 2 (quadrilateral *AOCD* − triangle *OCD*)

= 2 (triangle *AOC* + triangle *ACD* − triangle *OCD*)

= 2 triangle *AOC* + 2 triangle *ACD* − 2 triangle *OCD*

= 2 triangle *AOC* + parallelogram *CB* − parallelogram *CM*

= 2 triangle *AOC* + parallelogram *BN*

= 2 triangle *AOC* + 2 triangle *AOB*,

which proves the proposition.

Obs. From this and the preceding article we see, that the algebraic sum of the moments of any two forces, acting in one plane, about any point in that plane, is equal to the moment of their resultant about that point.

81.   *The algebraic sum of the moments of two forces, acting in one plane, and meeting in a point, about any point in the line of action of the resultant is zero.*

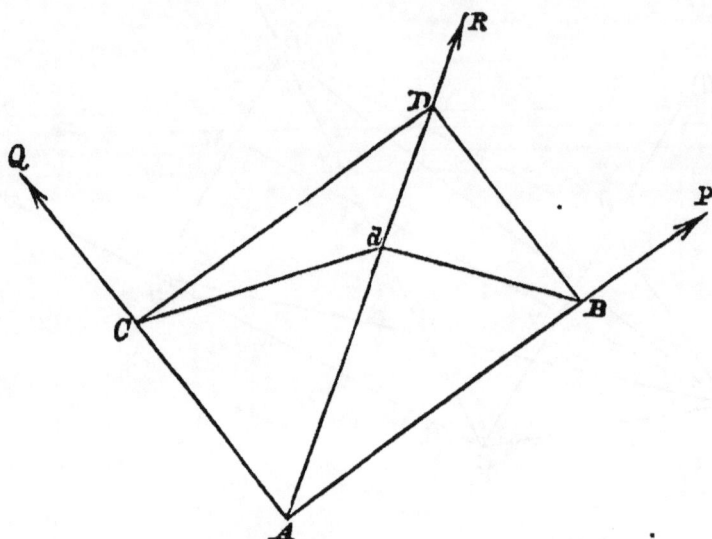

Let AB, AC represent tho two forces, P, Q.

Complete tho parallelogram $ABDC$.

First, to show that tho sum of the moments of P and Q about the point D is zero.

moment of P about $D = 2$ triangle ADB,

moment of Q about $D = 2$ triangle ADC,

and tho triangles ADB, ADC are equal;

$\therefore$ moment of P about $D =$ moment of Q about D,

and tho moments of P and Q are respectively *positive* and *negative;*

$\therefore$ tho algebraic sum of the moments of P and Q about D is zero.

Next let d be any point in the line of action of R.

Now   triangle AdB : triangle $ADB = Ad : AD$,

and   triangle AdC : triangle $ADC = Ad : AD$:

∴ triangle *AdB* : triangle *ADB* = triangle *AdC* : triangle *ADC*;
and ∴, since the triangles *ADB* and *ADC* are equal,

triangle *AdB* = triangle *AdC*;

i.e. moment of *P* about *d* = moment of *Q* about *d*;
and the moments of *P* and *Q* are respectively positive and
negative,

∴ algebraic sum of moments of *P* and *Q* about *d* is zero.

82. To shew that the moments of two *parallel* forces
about any point in the line of action of their resultant are
equal in magnitude and opposite in direction, we make use
of the figures in Article 79, and taking moments of *P* and *Q*
round any point *c* in the line of action of the resultant *R*, we
observe

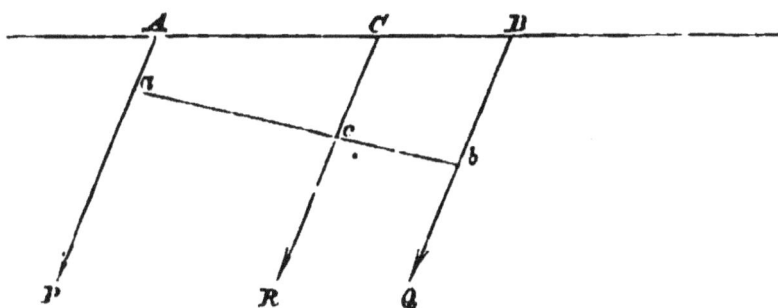

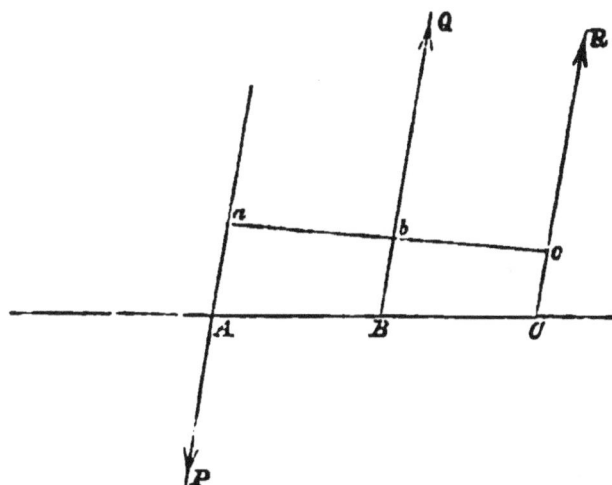

(1)   Since $P.ac = Q.bc$,

the moments of *P* and *Q* are equal in magnitude.

(2)   Since *P* and *Q* act in contrary directions with
respect to their tendency to turn *acb* in Case I. and *abc* in
Case II., regarded as rods moveable round the point *c*,

NOTE. We can readily extend the propositions proved in the preceding articles to any number of forces in one plane. For since the sum of the moments of two forces is equal to the moment of their resultant, we may substitute the resultant for the two forces; we may now combine this resultant with a third, and so on for any number of forces.

Hence we obtain the following conclusion:

The moment of the resultant of any number of forces in one plane, taken with respect to any point in that plane, is equal to the algebraic sum of the moments of the several forces with respect to the same point.

CHAPTER X.

OF MECHANICAL INSTRUMENTS.

83. A MECHANICAL Instrument, or Machine, is a contrivance for making a force, which is applied at one point, available at some other point.

The Simplest Machines are rods used in pushing, and ropes used in pulling, but what are called The Simple Machines, or Mechanical Powers, are

1. The Lever.
2. The Wheel and Axle.
3. The Pulley.
4. The Inclined Plane.
5. The Screw.
6. The Wedge.

84. THE LEVER.

We have already defined a Lever as a rigid rod, capable of turning round a fixed point in the rod, called the Fulcrum.

In the simplest cases, in which the Lever is employed as a Machine, three forces are brought into play:

1. The POWER applied to raise a weight or to overcome an obstacle.

2. The WEIGHT or obstacle to be overcome.

3. The REACTION of the fixed point or fulcrum.

85. Levers may be divided into three classes, according to the relative position of the points, where the *power* and *weight* are applied, with respect to the fulcrum.

In levers of the *first class*, the power and weight are applied on *opposite* sides of the fulcrum *C*, and act in the same direction.

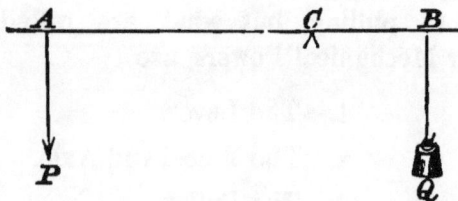

EXAMPLES. A poker between the bars of a grate, raising the coals. A spade. A pair of scissors is a double lever of this class, the rivet being the fulcrum.

In levers of the *second class*, the power and weight are applied on the *same side* of the fulcrum, and act in opposite directions, the power being applied at a *greater* distance from the fulcrum than the weight is.

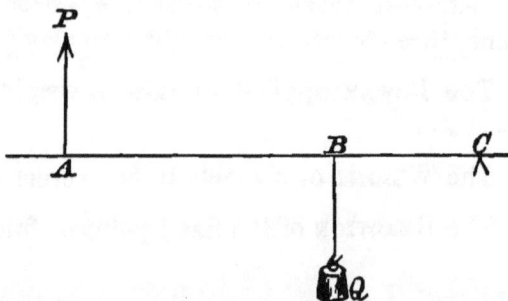

EXAMPLES. A wheelbarrow: the fulcrum being the point where the wheel presses on the ground. The chipping knife, in which the fulcrum is at the end attached to a bench, and the weight is the resistance of the substance to be cut, acting *upwards*, the power being applied by the hand *downwards*. This instrument is used for cutting turnips, sugar and tobacco. Another instance is an oar, the blade of the oar in the water being the fulcrum. A pair of nut-crackers is a double lever of this class

In levers of the *third class*, the power and weight are applied on the *same side* of the fulcrum, and act in opposite directions, the power being *nearer* to the fulcrum than the weight is.

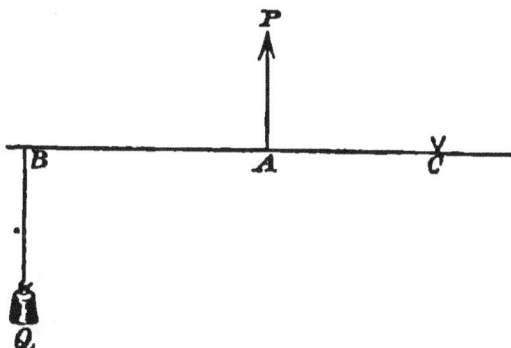

EXAMPLES. A man lifting a long ladder with one end resting on the ground. A pair of tongs is a double lever of this class.

86. *Conditions of Equilibrium of a Lever.*

Let F_1 and F_2 be the measures of two forces acting on a weightless lever, and let D_1 and D_2 be the measures of the perpendiculars drawn from the fulcrum to meet the lines of direction of F_1 and F_2.

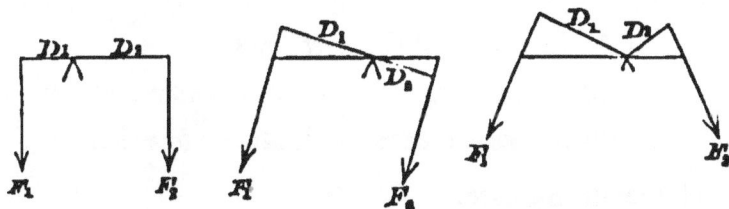

Then, since the lever is kept in equilibrium by the three forces F_1, F_2 and R, the reaction of the fulcrum; taking moments about the fulcrum, we shall have, as the condition of equilibrium in all cases,

$$F_1 \times D_1 = F_2 \times D_2.$$

If more than two forces act on a lever, the sum of the moments of the forces, which tend to turn the lever in one direction, about the fulcrum, must be equal to the sum of the moments of the forces, which tend to turn the lever in the contrary direction, about the fulcrum.

On Mechanical advantage and disadvantage.

87. The force applied to a machine to set it in motion is (as we said in Art. 84) called the Power (P), and the resistance to be overcome is called the Weight (W).

In the propositions which we are now discussing we determine the value of the Power which would suffice to balance the Weight, and any increase in this value of P will enable us to work the machine.

The *efficiency* or *working power* of a machine is measured by the fraction $\dfrac{W}{P}$, which is often called *The Modulus* of the machine.

When W is greater than P, the machine is said to work at *a mechanical advantage*, and when W is less than P, at *a mechanical disadvantage*.

To illustrate this from the cases of the straight and weightless Lever, acted on by two forces, in directions perpendicular to the Lever, we refer to the diagrams of Art. 85, and observe

(1) In a lever of the *first* class,

P will be *less* than W, if AC be *greater* than BC.

P will be *greater* than W, if AC be *less* than BC.

Hence, in this case,
mechanical advantage is the result, if P be further from the fulcrum than W,

mechanical disadvantage is the result, if P be nearer to the fulcrum than W.

(2) In a lever of the *second* class P is always less than W. Hence in this case mechanical advantage is always gained.

(3) In a lever of the *third* class P is always greater than W. Hence in this case mechanical disadvantage is always the result.

88. *If two weights be in equilibrium on a straight heavy lever in any position, which is not vertical, they will be in equilibrium in every position of the lever.*

Let P and Q be the weights suspended from the points A, B of the lever, whose fulcrum is C, centre of gravity G, and weight W.

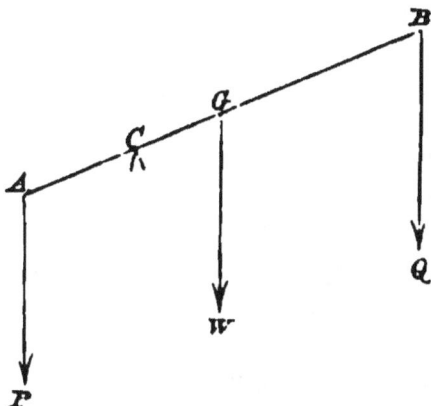

Then, since P, Q, W are parallel forces, acting always at the same points in the lever, their resultant will always pass through the same point of the lever. Now, since in *cne* position the lever balances about C, the resultant of P, Q, W, passes through C, and therefore in *all* positions of the lever the resultant will pass through C, and will therefore be counteracted by the resistance of the fixed point. Hence the lever will balance in every position.

EXAMPLES.—VII.

1. Pressures of 4 lbs. and 6 lbs. act, at points A and B, perpendicularly to a straight lever AB, 5 feet long. Find the position of D, the fulcrum and the pressure on it, when the pressures act in the same direction.

2. In a lever of the first class the arms are as 7 : 9. The pressure on the fulcrum is 36 lbs. Find the weights.

3. A uniform rod, 15 feet long, balances on a point 4 feet from one end, when a weight of 3 lbs. is suspended from that end. Find the weight of the rod.

4. ABC is a straight lever; the length of $AB = 7$ inches. that of $BC = 3$ inches; weights of 6 lbs. and 10 lbs. hang at A and B, and an upward pressure of 6 lbs. acts at C. Find the position of a fulcrum, about which the lever, so acted upon, would balance, and determine the pressure on it.

5.   A heavy rod balances itself on a point, one-third of its length from one end: if the rod be carried by two men, one at each end, what part of the weight will be supported by each?

6.   In a lever of the first class, if the distance of the fulcrum from one of the forces be a third of the whole length of the lever: shew that, when the direction of either of the forces is reversed, the fulcrum must then be placed at three times its former distance from the same force.

7.   A straight uniform lever, whose weight is 50 lbs. and length 6 ft., rests in equilibrium on a fulcrum, when a weight of 10 lbs. is suspended from one extremity.   Find the position of the fulcrum, and the pressure upon it.

8.   It is observed that a beam AB, whose length is 12 feet, will balance at a point 2 feet from the end A, but when a weight of 100 lbs. is hung from the end B, it balances at a point 2 feet from that end.   Find the weight of the beam.

9.   If a force of $2\frac{1}{2}$ lbs., acting vertically upwards at one extremity of a heavy rod, of length 4 feet, can just support $2\frac{2}{3}$ lbs., suspended at a distance of $1\frac{1}{2}$ feet from the other extremity, about which the rod can turn, find the weight of the rod and the pressure on the hinge.

10.   A rod, 6 feet long, is laid horizontally across two pegs, which are four feet apart, so that each end projects one foot beyond the pegs: weights of 3 lbs. and 5 lbs. being hung at the ends, required the pressures on the pegs.

Also, if an upward pressure be applied to the rod, at a point 2 feet from the more heavily laden end, in consequence of which the pegs now bear each the same pressure, determine the amount of pressure applied.

11.   Suppose weights of 2, 7, 5 and 3 lbs. to be suspended from a straight rod, at successive distances of 2, 2 and 13 inches from each other.   Where must a fulcrum be placed, for the rod to balance?

12. How would the mechanical advantage of an oar be modified by lengthening that part of it which is within the rowlock?

13. A heavy uniform bar, 10 feet long and of given weight W, is laid over two props in the same horizontal line, so that 1 foot of its length projects over one of the props. What must be the distance between the props, that the pressure on one may be double that on the other?

14. If the weights on a lever be 8 lbs. and 7 lbs., and the arms 8 inches and 9 inches respectively, at what point must a force of 1 lb. be applied perpendicularly to the lever, in order to keep them at rest?

15. Two weights, each equal to 8 lbs., hanging on a straight lever at points 12 inches and 18 inches from the fulcrum, and on the same side of it, are balanced by a single vertical force, acting at a point 16 inches from the fulcrum. Find the magnitude of the force, and show whether it acts at mechanical advantage or disadvantage.

16. Two men carry a uniform beam, 6 feet in length and weighing 16 stone, upon their shoulders, and at two feet from one end a weight of 4 stone is placed; what weight does each sustain, supposing the ends of the beam to rest on their shoulders?

17. Two men support a uniform heavy beam on their shoulders, which are at distances a and b from the ends; if the pressure on one man be r times that on the other, find the length of the beam.

18. Two uniform cylinders, whose lengths are 3 feet and 5 feet, and weights 15 lbs. and 9 lbs. respectively, are fastened together, so as to have their axes in the same straight line. Find the point about which they would balance.

19. A uniform beam, whose weight is 130 lbs. and length 12 feet, has weights of 15 lbs. and 25 lbs. hung on its ends. Find the point on which the system will balance.

20. A uniform rod rests with its extremities on two props. Find at what point a weight, equal to the weight of the rod, must be hung, in order that the pressures on the props may be in the ratio 2 : 1

21. The centre of gravity of a ladder, 40 feet long, is 15 feet from the thicker end. If the weight of the ladder be 152 lbs., find the part of the weight supported by each of two men, who carry the ladder, with one foot of its length projecting beyond the shoulder of each.

22. A beam, 24 feet in length, balances on a prop, 10 feet from the greater end. When the middle of the beam is placed on the prop, it requires a weight of 200 lbs. at the extremity of the lesser end, and also a weight of 20 lbs. at a distance of 4 feet from this end, to balance the beam. What is the weight of the beam?

23. A uniform beam, whose weight is ▓▓ lbs. and length 20 feet, has weights of 3, 5, 7, 9 lbs. hung on at distances 2, 4, 6, 8 feet respectively from one end. Find the point on which the system will balance.

24. (i) There is a heavy lever, one point of which (not the centre of gravity) is fixed. It balances when a weight P is suspended from one end. If nP be suspended at that end, find where P must be placed to balance it.

(ii) If the distance of the fulcrum from the nearer end be one tenth of its distance from the further end, find the greatest value of n.

25. A heavy uniform lever is 4 feet long, and weighs 2 lbs.; to one end is attached a weight of 3 lbs., to the other end a weight of 4 lbs.: find the smallest weight which can be attached to the lever, so that it may be in equilibrium, if the fulcrum be one foot from that end, to which the weight of 3 lbs. is attached.

26. A bent lever consists of two straight rods, of the same uniform material and thickness; the length of one rod is three times that of the other; what weight must be attached to the middle point of the shorter one, that the rods may be equally inclined to the horizon?

27. A uniform rod AB is kept at rest by a prop placed 2 inches from the end A and a force of $5\frac{2}{3}$ lbs. acting vertically upwards at B. The fulcrum is shifted 1 inch further from A, and, when a weight of 1 lb. is hung on to the middle point of the rod, it is found that a force of 5 lbs acting upwards at B is sufficient for equilibrium. Find the length and weight of the rod.

28. A uniform rod, to one end of which a weight of 5 lbs. is attached, will balance at a point 2 ft. from that end. When an additional weight of 6 lbs. is attached, it will balance at a point 1 ft. from that end. Find the length and weight of the rod.

29. *AB* is a lever 15 ft. long, moveable round a fulcrum at a distance of 4 ft. from *B*. A weight of 12 lbs. being suspended from *A*, and a weight of 8 lbs. from *B*, what force must act upwards at the middle point of *AB* to preserve equilibrium?

30. If weights of 3, 5 and 7 lbs. be suspended, at distances of 2, 4 and 8 feet from the fulcrum, on one arm of a straight lever, and weights of 5, 7 and 9 lbs. be suspended, at distances of 3, 6 and 7 feet from the fulcrum, on the other arm, where must a weight of 1 lb. be placed, so as to keep the lever at rest?

31. The arms of a bent lever are 3 and 5 feet. A force of 6 lbs. acting perpendicularly at the end of the shorter arm, is balanced by a force, acting at an angle of 30° to the longer arm at its end. Find this force.

32. In a straight lever of the second class, a force of 10 lbs. inclined at an angle of 60° to the lever, acts at one end, and balances a weight of 20 lbs. hanging 4 feet from the fulcrum. Find the length of the lever.

33. Two straight rods without weight, each four feet long, are loaded with weights 1 lb., 3 lbs., 5 lbs., 7 lbs., and 9 lbs., placed in order a foot apart. Show how to place one of the rods upon the other, so that both may balance about a fulcrum at the middle point of the lower: and explain how this may be done in two ways.

34. Two forces of 2 lbs. and 4 lbs. act at the same point of a straight lever, on opposite sides of it, the less one being perpendicular to the lever, and keep it at rest. Determine the direction of the greater force, and the pressure upon the fulcrum.

35. A bent lever consists of two uniform heavy beams whose lengths are as 1 : 2. Find the weight, which must be attached to the extremity of the shorter, in order that the arms may make equal angles with the horizon.

36. A bent lever ABC, having B for its fulcrum, and the angle ABC between its arms a right angle, has equal weights suspended from A and C, its extremities. Shew that, in the position of equilibrium, either arm is equally inclined to a vertical line, and to the line AC.

37. The arms of a lever are inclined to each other; shew that the lever will be in equilibrium with equal weights suspended from its extremities, if the point, midway between the extremities, be vertically above the fulcrum.

38. ABC is a weightless triangle having a right angle BAC, and $AB = 2 . AC$; if the triangle be suspended from A, and two weights P, Q, hanging at B, C, keep it at rest, with the sides AB, AC equally inclined to the vertical, find the ratio of P to Q.

39. The arms of a bent lever, of length 18 inches and 12 inches, are inclined to the horizon at angles of 45° and 30° respectively. If the greater weight be 16 lbs., find the less.

40. $ABCD$ is a quadrilateral figure, having the angles at B and D equal. Forces acting in AB and AD, tend to turn the figure about C, which is supported. Show that, when there is equilibrium, the forces are proportional to the opposite sides.

THE COMMON OR ROMAN STEELYARD.

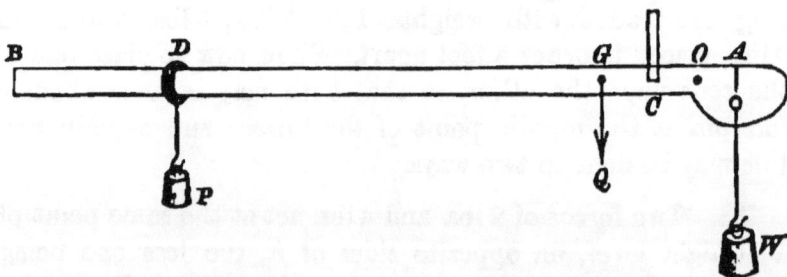

89. This balance consists of a straight lever AB suspended by the point C, and capable of turning round this point.

At the point A in the shorter arm is attached a hook, from which is suspended the substance, whose weight W is required.

A ring D, carrying a weight P of constant magnitude, can slide along the graduated arm CB, till P and W balance each other about C, when the lever is horizontal. The graduation, at which P rests, when this is the case, indicates the weight of the substance. In graduating the arm BC, account must be taken of the weight of the lever: let Q be the weight of the lever and G its centre of gravity, D the point from which P is suspended when it balances W at A: then taking moments about C, we have

$$P \cdot CD + Q \cdot CG = W \cdot CA \dots\dots\dots\dots\dots\dots(a).$$

If on the arm CA we take a point O, such that

$$P \cdot CO = Q \cdot CG,$$

the equation (a) becomes

$$P \cdot CD + P \cdot CO = W \cdot CA,$$

or

$$P(CD + CO) = W \cdot CA,$$

or

$$P \cdot OD = W \cdot CA;$$

$$\therefore\ OD : CA = W : P.$$

Now we may graduate OB by taking distances, measured from O, successively equal to CA, $2CA$, $3CA$, and marking them 1, 2, 3......and so on.

When P rests at the first of these graduations,

$$OD = CA \text{ and } \therefore W = P.$$

When P rests at the second of these graduations,

$$OD = 2CA \text{ and } \therefore W = 2P,$$

and so on.

Thus, when the weight of P is known, the weight of W is known also.

THE COMMON BALANCE.

90. This balance consists of a lever AB, called *the beam*, suspended from a fulcrum C, about which it can turn freely; the point C is a little above the centre of gravity G of the beam, and from the extremities A, B of the arms GA, GB, which ought to be similar and equal, are suspended two

scale-pans, in one of which is placed the substance, whose
weight W is required, and weights of known magnitude are
placed in the other, till their sum P just balances W; this
being the case, when the beam is exactly horizontal in a

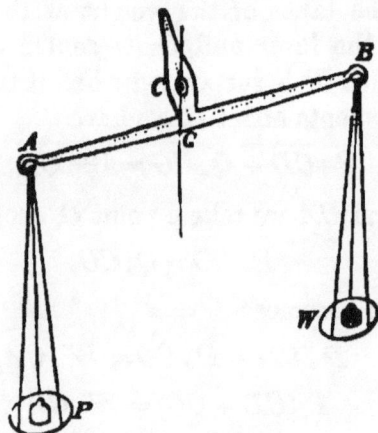

position of rest.   In this case, if the arms be perfectly equal
and similar, and the scale-pans also of equal weight, P will be
exactly equal to W.   If these weights differ by ever so little,
the horizontality of the beam will be disturbed, and, after
oscillating for a time, it will rest in a position inclined to the
horizon, and the greater this inclination is for a given
difference of P and W, the greater is the *sensibility* of the
balance.

91.   *The Requisites for a good balance.*

(1)   The balance ought to be *true;* i.e. the beam should
be horizontal when loaded with equal weights in the scales at
A and B.   This will be the case, if the scales be of equal
weight, and if the line drawn through C at right angles to
AB divide the beam into two equal and similar arms.

(2)   The balance ought to be *sensible;* i.e. the angle,
which the beam makes with the horizon ought to be easily
perceptible, when the weights P and W differ by a very
small quantity.

(3)   The balance ought to be *stable;* i.e. if the equilibrium be a little disturbed either way, there ought to be a
decided and rapid tendency to return to the original position
of rest, so as to ensure speed in the performance of a weighing.

The comparative importance of these qualities in a balance will depend upon the service, for which it is intended.

For weighing heavy goods, *stability* is of more importance.

For weighings requiring great accuracy, as in practical chemistry, *sensibility* is the quality desired.

A simple way of testing the accuracy of a balance is by interchanging P and W in the scales. The balance ought to retain the same position, when this is done.

THE DANISH STEELYARD.

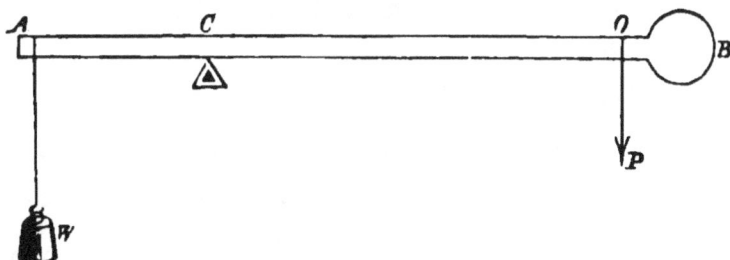

92. This instrument consists of a bar AB, terminating in a ball B, which serves as the *power*, and the substance to be weighed is suspended from the end A; the fulcrum C, which is generally a loop at the end of a string, being moved along AB, till P and W balance.

To graduate the instrument.

Let P be the weight of the steelyard, acting at O the centre of gravity: and let C be the position of the fulcrum, when P and W balance.

Then there is equilibrium, when
$$W : P = OC : AC.$$

Now if $W = P$, $OC = AC$, and $\therefore AC = \dfrac{AO}{2}$,

if $W = 2P$, $OC = 2AC$, and $\therefore AC = \dfrac{AO}{3}$,

if $W = 3P$, $OC = 3AC$, and $\therefore AC = \dfrac{AO}{4}$

And thus the graduations may be determined.

EXAMPLES—VIII.

1. A steelyard is formed out of a uniform bar, 38 inches long and weighing 12 lbs., the fulcrum being placed 5 inches from one end. If the moveable weight be $1\frac{1}{2}$ lbs., find the greatest weight, which can be determined by means of the instrument.

2. If the moveable weight, for which a common steelyard is constructed, be 1 lb., and a tradesman substitute a weight of 2 lbs., using the same graduations, shew that he defrauds his customers, if the centre of gravity of the steelyard be in the longer arm, and himself, if it be in the shorter arm.

3 If the common steelyard consist of a uniform rod, whose weight is $\frac{1}{p}$ of the moveable weight, and the fulcrum be $\frac{1}{4}$ of the length of the rod from one end; shew that the greatest weight that can be weighed is $\frac{3p+1}{p}$ times the moveable weight.

4. Where must be the centre of gravity of the common steelyard so that *any* moveable weight may be used with it?

5. If the common steelyard be correctly constructed for a moveable weight P, shew that it may be made a correctly constructed instrument for a moveable weight nP, by suspending at the centre of gravity of the steelyard a weight equal to $n-1$ times the weight of the steelyard.

6. The arms of a balance are in the ratio of 19 : 20. The pan, in which the weights are placed, is suspended from the longer arm. What is the real weight of a body, which apparently weighs 38 lbs.?

7. A body, the weight of which is 1 lb., appears to weigh 14 ounces, when it is placed in one scale of a false balance. What will be its apparent weight, when placed in the other scale?

8. One pound is weighed at each end of a false balance, and the sum of the apparent weights is $2\frac{1}{8}$ lbs., what is the ratio of the lengths of the arms

9. If a balance be false, having its arms unequal and in the ratio of 15 : 16, find how much per lb. a customer really pays for tea, which is sold to him from the longer arm at 3*s.* 9*d.* per lb.

10. If a Danish steelyard weigh *n* lbs., shew how to graduate it by ounces.

THE WHEEL AND AXLE.

93. This machine consists of a cylinder *HH'*, called *the axle*, and a *wheel AB*, the two having a common axis, terminating in pivots *C* and *C'*, about which the machine can turn; the pivots resting in fixed sockets at *C, C'*.

A rope, to one end of which the weight *W* is attached, passes round the axle, and has its other end fixed to the axle.

Another rope passes round the wheel, being attached at one end to the circumference of the wheel and at the other end the power *P* is applied.

The ropes pass round the *wheel* and *axle* in opposite directions, and thus tend to turn the machine in opposite directions.

The windlass and capstan are examples of the practical use of this mechanical instrument.

94.   *To find the condition of equilibrium on the Wheel and Axle.*

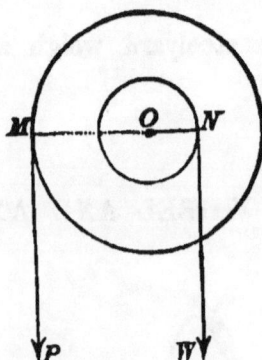

Suppose the Wheel and Axle to be cut by a vertical plane at the point of their junction, and that this figure represents the section.

We may then suppose P and W to act in this plane, and that their lines of action touch the circles at M and N.

From O the common centre draw OM and ON.

Then OM and ON, being drawn from the centre of the circles to the points of contact, will be perpendicular to PM and WN.

The axis of the machine being at rest, we may consider the machine as a lever, moveable round O as a fulcrum.

Then there will be equilibrium, when

$$P : W = ON : OM,$$

i. c. when $P : W = $ radius of axle : radius of wheel.

NOTE. If the thickness of the ropes cannot be neglected, we must suppose P and W to act along the middle of the ropes, and in this case

$$ON = \text{radius of axle} + \text{radius of rope},$$

$$OM = \text{radius of wheel} + \text{radius of rope}.$$

Examples.—IX.

1. If the radius of the wheel be 3 feet, the weight 18 lbs., and the power 3 lbs., what must be the radius of the axle?

2. If the radius of the wheel be 6 feet, the radius of the axle 2 feet, the weight 3 lbs., what must be the power, in order to produce equilibrium?

3. If the radius of the axle be 3 feet, and the radius of the wheel 9 feet, what power will be necessary, in order to keep a weight of 12 lbs. in equilibrium?

4. Explain how the "capstan" possesses mechanical advantage, and if the radius of the axle be 2 feet, and 6 men push, each with a force of 1 cwt., on spokes 5 feet long, find the weight they will just be able to support.

5. In what way must the power act so that the pressure on the axle may be the least possible?

6. If the string, to which the weight is attached, be coiled in the usual manner round the axle, but the string, by which the power is applied, be nailed to a point in the rim of the wheel, find the position of equilibrium, the power and weight being equal.

7. The radius of the wheel being three times that of the axle, and the string on the wheel being only strong enough to support a tension equivalent to 36 lbs., find the greatest weight which can be lifted.

8. If the axis of the axle do not coincide with the centre of the wheel, shew that the sum of the weights, which a given power will support in the two positions, in which the diameter of the wheel containing the centre of the axle is horizontal, is double the weight, which it would support, if the machine were accurately made.

9. Compare the greatest and least weights which can be supported by a force acting on a wheel with a square axle.

THE PULLEY.

95. The *pulley* is a small circular disc or wheel, having a uniform groove cut on its outer edge, and it can turn freely about an axis, which passes through its centre. This axis rests in sockets within the *block*, to which the pulley is attached.

When the block is fixed, the pulley is said to be fixed ; in other cases it is moveable. A cord passes over the pulley along the groove, and at its extremities the power and weight are applied.

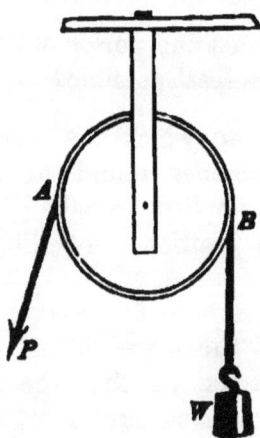

The pulley is very useful for changing the direction of a string; and, assuming that the tension of a string is not altered by passing over a small pulley, the tension at all points of the string between the points of application of P and W will be the same.

When the pulley is fixed, no mechanical advantage is gained by its use, beyond that of greater convenience in applying the force.

96. *To find the conditions of equilibrium on a single moveable pulley.*

(1) When the strings are parallel.

Let the string $PABC$ have one extremity fixed at C, and, after passing under the moveable pulley, suppose it to be passed over a fixed pulley D and pulled by a force P. The weight W is suspended by a string whose direction is in a line with O the centre of the moveable pulley.

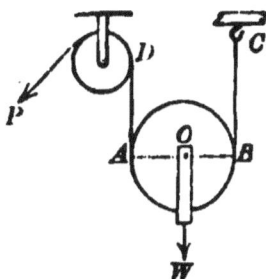

Then the tension of each of the strings DA, CB is P.

Hence the pulley is acted on by three parallel forces P, P and W; and therefore, when there is equilibrium,

$$W = 2\,P.$$

Obs. If weight of pulley (w) be taken into account, $2P = W + w$.

(ii) When the strings are not parallel.

Let the string quit the pulley at A and B.

Then, since the tension along AP is equal to that along BC, their resultant will bisect the angle between them, Art. 22, and this resultant must be equal and opposite to the weight W suspended from the axis of the pulley, and acting in a vertical direction.

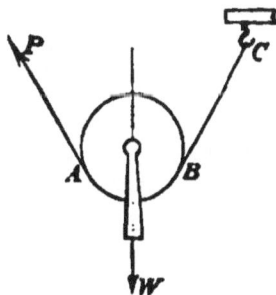

Hence AP, BC must be equally inclined to the vertical. Let θ be this inclination. Then the resultant of the two tensions, which we may regard as acting at A and B, is

$$2P.\cos\theta,$$

and this must be equal to W;

$$\therefore 2P.\cos\theta = W$$

is the condition of equilibrium.

97. *To find the condition of equilibrium for a system of pulleys, in which each pulley hangs by a separate string, the strings being parallel.*

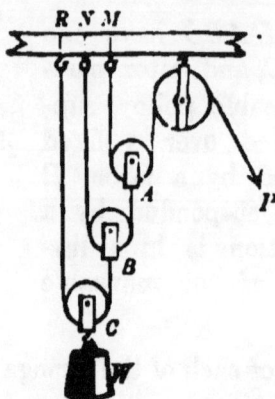

In this system a string, acted on by the power P, passes under the pulley A, and is fastened to the block at M.

A string, attached to A, passes under the pulley B, and is fastened to the block at N, and so on for any number of pulleys.

The weight W is suspended from the lowest pulley.

Then, since W is supported by the tension of the strings RC, BC,

$$BC\text{'s tension} = \frac{W}{2}.$$

Again, $$\text{tension of } AB = \frac{\text{tension of } BC}{2} = \frac{W}{4},$$

and $$\text{tension of } PA = \frac{\text{tension of } AB}{2} = \frac{W}{8}.$$

But $$\text{tension of } PA = P,$$

$$\therefore P = \frac{W}{8}.$$

Thus, when there are three pulleys, $P = \dfrac{W}{2^3}$,

and, similarly, when there are n pulleys, $P = \dfrac{W}{2^n}$,

$$\therefore P : W = 1 : 2^n.$$

NOTE. If the pulleys have weight, an additional force p will be required to assist P. Calling the weights of the pulleys, commencing with the *highest*, $w_1, w_2, w_3 \ldots \ldots w_n,$

$$p = \frac{w_1}{2} + \frac{w_2}{4} + \frac{w_3}{8} + \ldots \ldots \frac{w_n}{2^n},$$

the terms on the right-hand side of the equation being obtained by taking the formula $P = \dfrac{W}{2^n}$, and making $W = w_1,$ $w_2 \ldots w_n,$ successively, and $n = 1, 2 \ldots n,$ successively.

NOTE. This is usually called the *First System of Pulleys.*

98. *To find the condition of equilibrium for a system of pulleys, where there are two blocks and the same string passes round the pulleys.*

In this system of pulleys the *same* string passes round each of the pulleys as in the figure, and the parts of the string between successive pulleys are taken to be parallel.

The tension of the string is the same throughout, and is equal to P. Hence, if n be the number of strings at the lower block, nP will be the *resultant* of the upward tensions of the strings upon the lower block.

This resultant must be equal to W, when there is equilibrium,

that is, $\qquad nP = W$

is the condition required, which may be expressed thus,

$$P : W = 1 : n.$$

NOTE. This is usually called the *Second System of Pulleys.*

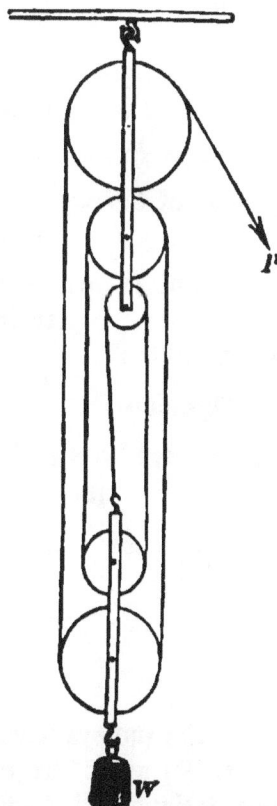

99. *To find the condition of equilibrium in a system of pulleys, in which all the strings are attached to the weight.*

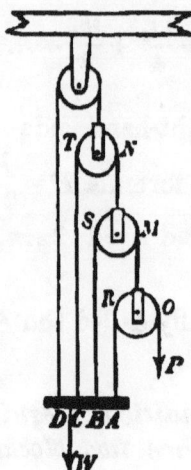

The figure represents the system.

The weight W is supported by the tension of the strings RA, SB, TC,......attached to a bar AD.   Now

tension of $RA = P$,

tension of $SB =$ tension of MO
$\qquad\qquad =$ tension of $RA +$ tension of $OP = 2P$,

tension of $TC =$ tension of NM
$\qquad\qquad =$ tension of $SB +$ tension of $MO = 2^2.P$, and
so on.

Therefore $W = P + 2P + 2^2.P + \ldots\ldots,$

and if there be n pulleys,

$$W = P + 2P + 2^2.P + \ldots\ldots + 2^{n-1}.P$$

$$= P(1 + 2 + 2^2 + \ldots\ldots + 2^{n-1})$$

$$= P.\left(\frac{2^n - 1}{2 - 1}\right), \quad \text{by Geometrical Progression,}$$

$$= P.(2^n - 1).$$

If the pulleys have weight, they all, except the uppermost, *assist* P; and if we call the assistance, which they afford, p, and designate the weights of the pulleys beginning with the *lowest* by $w_1, w_2, w_3, \ldots$

$p = (2^{n-1}-1)w_1 + (2^{n-2}-1)w_2 + \ldots + (2^2-1).w_{n-2} + (2-1).w_{n-1},$
the terms on the right-hand member of the equation being obtained by taking the formula

$$W = P(2^n - 1),$$

and making P equal $w_1, w_2 \ldots\ldots w_{n-2}, w_{n-1}$, successively,

the number of pulleys being $n-1, n-2 \ldots\ldots 2, 1$, successively.

NOTE. This is usually called the *Third System of Pulleys.*

EXAMPLES.—X.

1. In a single moveable pulley, if the strings be not parallel, and $P = W$, what must be the angle between the strings?

2. Show that no mechanical advantage is gained by the single moveable pulley, unless the weight of the pulley be less than the power.

3. If the angle between the strings of the single moveable pulley be two-thirds of a right angle, what must be the ratio of the Power to the Weight, in order to produce equilibrium?

FIRST SYSTEM.

4. In a system of three pulleys a weight of 8 lbs. is attached to the lowest pulley. Neglecting the weights of the pulleys, find the power necessary to produce equilibrium.

5. In a system of pulleys in which each pulley hangs by a separate string there are 3 pulleys of equal weights. The weight attached to the lowest is 32 lbs., and the power is 11 lbs. Find the weight of each pulley.

6. In a system of 3 pulleys a weight of 5 lbs. is attached to the lowest pulley. Supposing the weight of each pulley to be 3 lbs., find the force required to sustain equilibrium.

7. If there be 4 pulleys, whose weights, commencing with the highest, are 1, 2, 4, and 8 lbs. respectively, and W be 160 lbs., find P.

8.   If there be three pulleys, and the weight of each be 1 lb., find the force capable of supporting a weight of 9 lbs.

9.   If there be three pulleys, the weight of each being W, but no weight attached to the lowest, shew that there will be equilibrium, when $P : W :: 7 : 8$.

10.   If there be three pulleys of equal weight, what must be the weight of each, in order that a weight of 56 lbs. attached to the lowest may be supported by a power equal to 7 lbs. 14 oz.?

11.   What must be the weight of each of three pulleys that P may equal W, the pulleys being all of equal weight?

12.   If all the pulleys, except the lowest, be considered weightless, and the weight of the lowest and the power be each p lbs., and the weight attached be w lbs., shew that w is some odd multiple of p.

13.   If $P = 2$ lbs. in a system of 4 pulleys, each hanging by a separate string, and the weight of each pulley together with the string beneath it be 1 lb., shew that $W = 17$ lbs.

14.   If on the same system of pulleys, in which each pulley hangs by a separate string, P can support W, and P' causes a pressure W' on the beam, to which the strings are all attached, shew that $\dfrac{W}{P} - \dfrac{W'}{P'} = 1$.

15.   If the beam, to which the strings are attached, be fixed at one point only, about which it is capable of revolving, and there are only two moveable pulleys, find the position of the point, in order that the beam may remain horizontal.

SECOND SYSTEM.

16.   In the system of pulleys, in which the same string passes round all the pulleys and the parts of it between the pulleys are vertical, if there be three pulleys at the lower block, and this block weigh 8 lbs., find the power which will support 1 cwt.

17. In the system of pulleys, in which the same string passes round all the pulleys, if $P = 2$ lbs., the lower block weigh 8 lbs., and contain 3 pulleys, and the string be fastened to the lower block, shew that $W = 6$ lbs.

18. A man supports a weight, equal to half his own weight, by a system of pulleys, in which the same string passes round all the pulleys, the upper block being attached to the ceiling: if there be 7 strings at the lower block, find his pressure on the floor, on which he stands.

19. What weight will be supported, if there be 3 pulleys in the lower block, the string being fastened to the upper block, and the weight of the lower block being equal to 3 times the power?

20. Supposing that a power of 3 lbs. will just support a weight of 10 lbs. suspended from the lower block, the number of strings being 4, what is the weight of the lower block?

21. If the weight of the lower block and the power be each p lbs., and the weight attached to the lower block be w lbs., shew that w is some odd or even multiple of p, according as the end of the string is fastened to the upper or lower block.

Third System.

22. A power P and a weight W are in equilibrium on a system of pulleys, in which all the strings are parallel and attached to a uniform bar, from which the weight is suspended, the weights of the pulleys being neglected. If the number of pulleys be three, and the strings be equidistant, from what point of the bar ought the weight to be suspended, that the bar may rest in a horizontal position?

23. In a system of 6 pulleys of equal weight, where each pulley is attached to a string, which is attached also to the weight, find the ratio, which the weight of each pulley must bear to the weight supported, in order that there may be equilibrium, without any power being applied.

THE INCLINED PLANE.

100.　By an inclined plane, as a mechanical instrument, is meant a plane inclined to the horizon.

The figure represents a section of the inclined plane, made by a *vertical* plane perpendicular to the inclined plane.

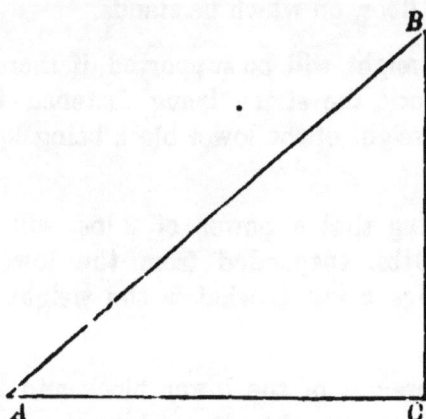

AB is called *the length* of the plane.

BC, which is taken to be perpendicular to *AC*, is called *the height* of the plane.

AC is called *the base* of the plane.

The angle *BAC* is called *the inclination* of the plane.

When a body is in contact with a *smooth* plane, there is a mutual action between the body and the plane, acting *at right angles* to the plane. The force thus brought to bear on the body is called the *reaction* of the plane, and the reason for this reaction being equal to the pressure of the body on the plane is to be explained thus:—

Reaction is always contrary and equal to action: or, the mutual actions of two bodies upon each other are always equal, and directed towards opposite parts. Whatever draws or presses another, is as much drawn and pressed by that other. If any one press a stone with his finger, his finger is also pressed by the stone. If a horse draw a stone tied to a rope, the horse will be equally drawn back towards the stone.

101. *To find the condition of equilibrium on a smooth Inclined Plane, when the Power acts parallel to the plane.*

Let a body, D, whose weight is W, be pulled by a force P acting parallel to the plane, and let the body be at rest.

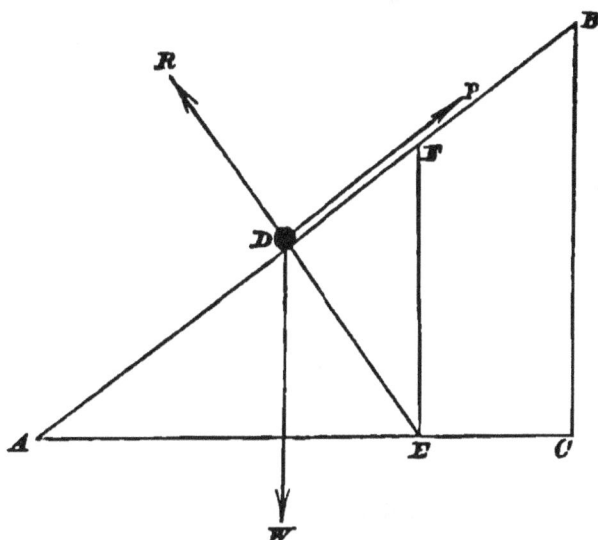

The body is acted upon by three forces :

P, the power, acting parallel to AB,

W, the weight, acting parallel to BC, i.e. vertically,

R, the reaction of the plane, acting at right angles to AB.

Produce RD to meet AC in E.

From E draw EF parallel to BC.

Then, since the three sides of the triangle DFE are *parallel* to P, W, R respectively, the sides taken in proper order are also *proportional* to P, W, R, by Art. 38;

$$\therefore P : W = DF : FE.$$

Again, EFD, ABC are similar triangles, for

the right angle $FDE =$ the right angle BCA,

and the angle $EFD =$ the angle ABC,

and $\therefore$ the remaining angles FED, BAC are equal.

Hence $\qquad DF : FE = CB : BA;$

$\qquad$ and $\therefore P : W = CB : BA,$

that is $P : W =$ height of plane : length of plane.

Similarly we can prove that $P : R = CB : CA,$

$\qquad\qquad$ and $W : R = BA : CA.$

102.   *To find the condition of equilibrium on a smooth Inclined Plane, when the Power acts horizontally.*

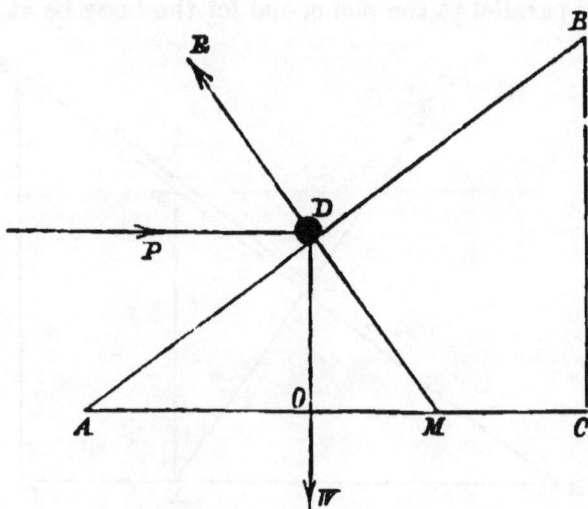

Let the body D be kept at rest by three forces:

P, the power, acting horizontally,

W, the weight of D, acting vertically downwards,

R, the reaction of the plane, acting at right angles to AB.

Produce RD to meet AC in M.

Now, angles MDO, ODA make up a right angle,
and angles DAO, ODA are together equal to a right angle;

$$\therefore \text{ angle } MDO = \text{angle } DAO;$$

that is,   angle $MDO =$ angle BAC.

Also,   angle $MOD =$ angle BCA;

$$\therefore MDO, BAC \text{ are similar triangles.}$$

Then, since the sides of the triangle MOD are parallel, and therefore proportional, to P, W, R,

$$P : W = MO : DO$$
$$= BC : AC.$$

Also $P : R = BC : AB,$

ciii. By the aid of Trigonometry, we can find the condition of equilibrium on a smooth inclined plane in a more general form.

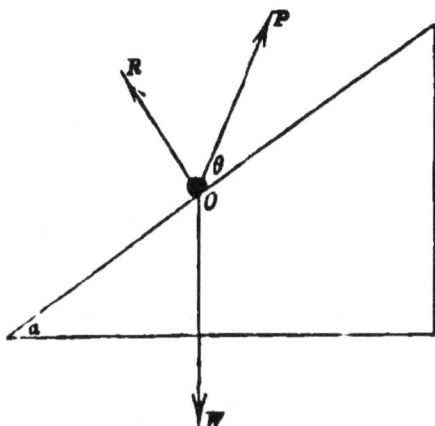

Let a body O, whose weight is W, be supported on a smooth inclined plane by a force P, the direction of which makes an angle θ with the plane.

Let a be the inclination of the plane.

Then the forces acting on the body O are

W, the weight of the body, acting vertically downwards,

R, the reaction of the plane, acting at right angles to the plane,

P, the given force.

Then, since there is equilibrium, we have by Art. xxxix.
$$P : W : R = \sin ROW : \sin ROP : \sin POW,$$
$$= \sin (180° - a) : \sin (90° - \theta) : \sin (90° + a + \theta),$$
$$= \sin a : \cos \theta : \cos (a + \theta). \quad \text{(Trig. Art. 101, 99, 102.)}$$

Two particular cases are to be especially noticed:

(1) When P acts *parallel* to the plane, $\theta = 0$, $\cos \theta = 1$, and $\therefore P : W : R = \sin a : 1 : \cos a.$

(2) When P acts *horizontally*, $\theta = -a$, $\cos \theta = \cos a$, and $\cos (a + \theta) = 1$, (Trig. Art. 104, 67), and $\therefore P : W : R = \sin a : \cos a : 1.$

EXAMPLES.—XI.

INCLINED PLANE.

1. If W be 3 tons, find P, acting parallel to the plane, when the height of the plane is to its base as 5 : 12.

2. Find the pressure on the plane, when the height of the plane is to its base as 3 : 4, and the weight supported is 10 lbs., the power being parallel to the plane.

3. If, when P acts along the plane, $R : P = 3 : 4$, express R and P in terms of W.

4. Find the horizontal force necessary to support a body whose weight is 12 lbs. upon a plane, whose base is to its length as 4 : 5.

5. If the pressure on the plane be 2 lbs., and the power acting horizontally 1 lb., what is the weight? and what the inclination of the plane?

6. A force of 15 lbs., acting horizontally, supports a weight of 20 lbs. on an inclined plane: what force, acting along the plane, will support the same weight?

7. A railway train, weighing 50 tons, is drawn up an incline of 1 in 30 by a rope: what is the least strain on the rope, if 10 lbs. per ton be allowed for resistance to motion on level rails?

8. A force of 15 lbs., acting horizontally, supports a weight of $5\sqrt{3}$ lbs. on an inclined plane; find the inclination of the plane to the horizon.

9. The weight supported upon an inclined plane is $2\sqrt{2}$ lbs., and the plane is inclined at half a right angle to the horizon; find the power, which, acting along the plane, will support the weight.

10. A weight of 56 lbs. rests upon a smooth plane inclined at 45° to the horizon. What is the smallest horizontal force required to move it up the plane?

11. What force, acting horizontally, will sustain a weight of 12 lbs. on a plane inclined to the horizon at an angle of 60°?

12. What force, acting horizontally, will sustain a weight of 10 lbs. on a plane inclined to the horizon at an angle equal to half of one of the angles of an equilateral triangle?

13. If the force, which will support a weight, when acting parallel to the plane, be half that, which will do so, acting horizontally, find the inclination of the plane.

14. Equal weights are attached to the ends of a string; one of which rests on a plane inclined at 45° to the horizon, and the other hangs vertically over the summit of the plane, and rests on the ground beneath. Find the pressure of the latter on the ground.

15. What power, acting parallel to a smooth plane inclined at an angle of 30°, is necessary to sustain a weight of 4 lbs. on the plane?

16. Shew that, if P, instead of acting parallel to the plane, were to make the same angle with the vertical as the pressure of the plane on the body, the pressure on the plane would be equal to P.

17. Which will support the greater weight, a power acting horizontally, or the same power acting parallel to the plane?

18. A weight of 20 lbs. is supported by a string fastened to a point in an inclined plane, and the string is only just strong enough to support a weight of 10 lbs.; the inclination of the plane to the horizon being gradually increased, find when the string will break.

19. Two unequal weights W and W', connected by a string, are placed upon two smooth inclined planes, the string passing over the intersection of the planes. Find the ratio between the weights, when there is equilibrium.

20. The least power which will sustain a body on a certain inclined plane is sufficient, when acting horizontally, to support

a body weighing half as much again as the other on a plane, whose inclination is half that of the former: find the angle of inclination.

21. The powers required to keep a given weight at rest on an inclined plane, are 10 lbs. and 20 lbs., in the cases of the power acting along the plane and horizontally; find the inclination of the plane.

22. A weight of 20 lbs. is supported on an inclined plane by a power of 12 lbs., acting parallel to the plane. Shew that, if it were required to support the same weight on the same plane by a force applied horizontally, the force must be increased in the ratio of 4 to 5, whilst the pressure on the plane will be increased in the duplicate of the same ratio.

23. If the power, acting along an inclined plane, the pressure on the plane, and the weight be as $1 : \sqrt{3} : 2$, find the inclination of the plane to the horizon.

24. When a certain inclined plane ABC, whose length is AC, is placed upon AB as a base, a power of 3 lbs. can support on it a weight of 5 lbs.; what weight could the same power support, if the plane were placed on BC as base, so that AB is then the height of the plane?

25. If a plane be inclined to the horizontal at an angle a, and a force $W.\tan a$ act upwards, parallel to the plane, on a weight W which is in the plane, what force must act down the plane to keep the weight in equilibrium?

26. Give a geometrical construction for determining the direction, in which the power must act, when it is equal to the weight, and shew that, if R_1 be the pressure on the plane in this case, and R the pressure, when the power acts parallel to the plane, $R_1 = 2R$.

THE SCREW.

104. The screw is a spiral thread, running along the surface of a circular cylinder, which may be imagined to be generated thus:

Let AG be a rectangle, whose base AB is exactly equal to the circumference of the cylinder; make the rectangles BD, CF, EH... equal in every respect, and draw the straight lines AC, DE, FG...; then, if the rectangle BH be applied to the surface of the cylinder, so that the base AB coincides with the base of the cylinder, the broken lines AC, DE, FG... will form a continuous line on the surface of the cylinder, the point C coinciding with D, E with F, and so on. If we now suppose this *line* to become a protuberant thread, perpendicular to the plane of the rectangle, we obtain a *screw*, in which the distance between any point of one thread and the one next below it, measured parallel to the axis of the cylinder, is everywhere the same and equal to BC.

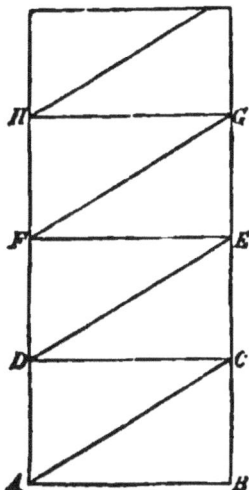

The angle CAB, which the thread at any point makes with the base of the cylinder, is called the *pitch* of the screw.

The screw formed on the *solid* cylinder, as above, works in a *hollow* cylinder of equal radius, in which a spiral groove is cut exactly equal and similar to the thread on the solid cylinder, and in which groove the thread of the solid screw can work freely.

A solid and hollow screw, related as above, are called *companion screws*; and, when in action, one of them is fixed and the other is turned by means of a lever, fixed into the cylinder at right angles to its axis. By turning the lever a weight is raised, or a pressure produced, at the end of the screw, which pressure acts in direction of the axis of the screw.

When the hollow screw is small, it is sometimes called a *nut*.

The annexed figure, representing the appearance of a solid screw, will assist the reader in understanding that a screw is nothing more than an inclined plane, constructed on the surface of a cylinder.

cv.   *To find the condition of equilibrium on the Screw.*

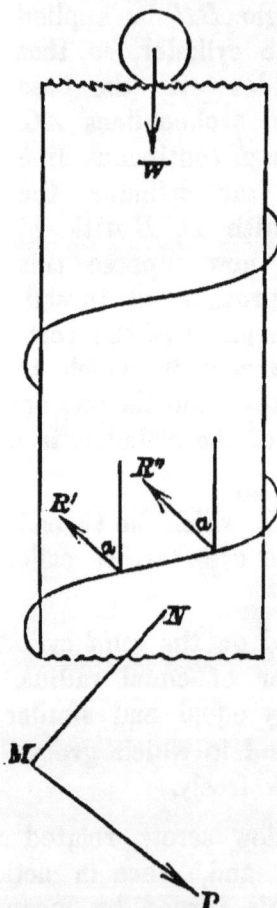

The forces acting on the screw will be

P, the power, acting horizontally at right angles to the rod *MN* (which is at right angles to the axis of the Screw, and whose length is *a*), and producing a vertical pressure upwards in direction of the axis;

W, a weight placed at the end of the screw and acting vertically downwards;

and a series of forces *R'*, *R''*... arising from the pressure of the hollow screw on each point of the solid screw, with which it is in contact.

The forces *R'*, *R''*... act at right angles to the thread, and will therefore make an angle *a*, equal to the *pitch* of the screw, with the lines drawn vertically from the points of contact.

Resolving *R'*, *R''*... vertically and horizontally, we shall have

R' . cos *a*, *R''*. cos *a*,... acting vertically upwards,

and *R*. sin *a*, *R''*. sin *a*,... acting horizontally, and tending to turn the screw in a direction contrary to that, in which *P* tends to turn it.

Also, each of these horizontal forces acts at an arm *r*, equal to the radius of the cylinder.

Hence, taking moments, $(R' + R'' + ...).\sin a . r = P . a...(1)$,
and $(R' + R'' + ...).\cos a = W......(2)$.

Dividing (1) by (2)

$$\tan a . r = \frac{P.a}{W},$$

or $$\frac{P}{W} = \frac{r.\tan a}{a}.$$

This condition of equilibrium may be expressed in another form, thus, since

$2\pi . r . \tan a$ = distance between two threads, measured parallel to the axis,

and $2\pi . a$ = circumference of the circle described by *M*,

$$\frac{P}{W} = \frac{2\pi . r . \tan a}{2\pi . a}$$

$$= \frac{\text{distance between two threads}}{\text{circumference of circle whose radius is } MN}.$$

THE WEDGE.

106. The Wedge is a solid triangular prism.

Its two ends are equal and similar triangles.

Its three sides are rectangular parallelograms.

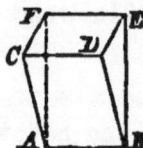

AB is called its *edge: CDEF* its *head.*

It is used for separating bodies, or parts of the same body, which adhere strongly to each other.

The edge of the Wedge is introduced into a small cleft, and it is then driven forward by blows of a hammer applied at its head.

The mode of working this machine is quite different in principle from the method used in the other machines, which have been described.

They are worked by the *regular* and *steady* application of a power, acting uniformly at that point of the machine, to which it is applied, and *gradually* producing motion: but in this machine the power is applied by *sudden impulses.*

Hence any investigation for finding the relation between the power and weight in this machine must involve considerations, which cannot be explained by the principles of Elementary Statics.

Hatchets, chisels, nails, carpenters' planes, swords, are modifications of the wedge.

ON FRICTION.

107. In our investigation of the Conditions of Equilibrium for the Mechanical Instruments we assumed that the surfaces under consideration were *perfectly smooth.* But, since in practice no surface is perfectly smooth, it is necessary, in applying the laws of equilibrium to particular problems, to take into consideration the resistance to motion, which is brought into play by an attempt to move a *rough* body over a *rough* surface. This resistance is called Friction.

108.   Friction is a force, of which the practical use may be seen from the following instances:

(1)   At every step we take in walking we bring friction into play.   If we attempt to walk on a surface approaching to perfect smoothness, as a polished oaken floor, or a sheet of ice, our feet have a tendency to slip, because but a slight amount of friction can be brought into action.

(2)   If a wedge be driven by a blow into a block of wood, but for the friction between the wedge and the block, the wedge would fly back.

109.   On attempting to displace a body at rest on a rough horizontal plane, we experience three kinds of resistance:

(1)   On trying to *lift* the body *off* the plane, we experience an opposition as it were in the body itself, arising from the attraction of the Earth.

(2)   On attempting to *press* the body *against* the plane, on which it rests, we find that the plane resists our effort.

(3)   On trying to *push* the body *along* the plane, we experience a resistance, varying according to the nature of the surfaces in contact.

This resistance is called Friction, and its laws are as follows:

(*a*)   Its *direction* is opposite to that of attempted motion.

(*β*)   Its *magnitude* is just sufficient to prevent motion, but no more than a certain amount can be called into play. If more than this amount be required to prevent motion, motion will ensue.   This greatest possible amount is termed *Limiting Friction*, and when this is just called into action, the body is said to be *in a state bordering on motion.*

110.   The *statical* laws of Limiting Friction are

I.   When the substances in contact remain the same, the Limiting Friction varies as the Pressure between the bodies.

II. The amount of Limiting Friction is independent of the area in contact.

If any oily matter be introduced between the substances, a smaller amount only of friction is capable of being called into play, that is, the Limiting Friction is then less.

All the Laws of Friction have been obtained by experiment.

111. If R be the pressure between two bodies in contact, and F the amount of Limiting Friction, then by Law I,

$$F \text{ varies as } R,$$

or
$$F = \mu R,$$

where μ is a constant quantity, to be determined by experiment.

112. When we say that μ is constant, we mean that it is independent of R, independent of the extent of surface in contact, and dependent only upon the nature of the surfaces.

μ is called *The Coefficient of Friction*.

cxiii. *If a body be on the point of motion, when placed on an inclined plane, inclined at an angle a to the horizon, then the coefficient of friction between the body and the inclined plane* = tan a.

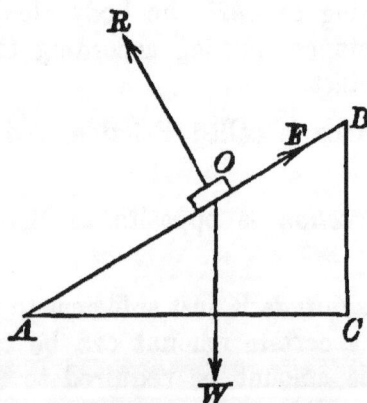

Let ABC be a rough plane, on which a body O when placed is in a state just bordering on motion.

Then the body is kept at rest by three forces,

W, the weight of the body,

R, the resistance of the plane,

F, the force of friction acting along the plane.

Hence, by Art. 101,

$$\frac{F}{R} = \frac{BC}{AC}.$$

Now, by Art. 111, $\qquad F = \mu R;$

$$\therefore \frac{\mu R}{R} = \frac{BC}{AC};$$

$$\therefore \mu = \frac{BC}{AC}.$$

Hence, if a be the inclination of the plane to the horizon,

$$\mu = \tan a.$$

MISCELLANEOUS EXERCISES.

1. If two equal forces, acting on a point, are balanced by two other equal forces acting at the same point, show that the latter two are equally inclined to the former two. Are all four forces necessarily equal ?

2. The resultant of two forces P and Q, acting on a particle, is the same, when their directions are inclined at an angle θ, as when they are inclined at an angle $\frac{\pi}{4} - \theta$ to each other: shew that $\tan \theta = \sqrt{2} - 1$.

3. If the actual lines of action of three forces, which keep a point at rest, be represented by lines drawn from the point of application, the greatest of the three will always be opposite to the least of the angles made by the lines, and the least force will be opposite the greatest angle.

4. Through any point P on the diagonal AC of a parallelogram $ABCD$ a line is drawn parallel to AD meeting AB in L, CD in N, and another parallel to AB meeting BC in M, AD in O. Prove that forces represented in magnitude and direction by ML, NO, AC will be in equilibrium.

5. Equal weights are supported on two inclined planes of equal length, but of different heights ; and the power required on the steeper plane is equal to the resistance exerted by the other : also, this resistance is three times the resistance of the steeper plane. Compare the two powers employed, and their relation to W.

6. A point is taken within or outside a triangle, and lines are drawn from it to the angular points of the triangle. Prove that the resultant of the forces represented by these lines is represented by 3 times the line joining the point and the centre of gravity of the triangle.

7. A line is drawn parallel to the base of an isosceles triangle, bisecting the perpendicular from the vertex on the base and meeting the sides. Shew that the distance of the centre of gravity of the quadrilateral figure thus formed from the vertex is seven-ninths of the perpendicular.

8. A weight of 25 lbs. hangs at rest, attached to the ends of two strings, the lengths of which are 3 and 4 feet, and the other ends of the strings are fastened at two points in a horizontal line, distant 5 feet from each other; find the tension of each string.

9. A uniform heavy rod is supported by two strings attached to its two ends and to a fixed point, so that the strings and the rod form an equilateral triangle. Compare the tension of the strings with the weight of the rod.

10. The resultant of two forces P and Q makes an angle of 30° with the direction of P, and is equal to $\sqrt{3} \cdot Q$; shew that P may be equal either to Q or to $2Q$.

11. The angle of inclination of an inclined plane being 30°, a force P acting horizontally keeps the weight in equilibrium. If P now act in a direction making an angle of 30° with the plane and above it, shew that it will still maintain equilibrium, and that the pressure on the plane will be reduced one-half.

12. Three forces, acting in the same plane, keep a point at rest, the angles between the directions of the forces being 135°, 120°, and 105°. Compare their magnitude.

13. A boat, moored by the bank of a rapid river, is also attached by a rope to a stake higher up the stream on the opposite bank. Shew that, if the boat be loosed from its moorings, it will swing round to the opposite bank.

14. If three forces, acting at a point, be represented in magnitude and direction by three straight lines drawn from

that point, then, if the forces are in equilibrium, the triangles formed by joining the extremities of the lines are equal.

15. If a point be acted upon by three forces, parallel and proportional to the three sides AB, BC, DC of a quadrilateral, shew that the resultant of the forces is represented in magnitude and direction by ECF, E being the middle point of AD, and CF being equal to EC.

16. Three pegs, A, B, C, are stuck in a wall in the angles of an equilateral triangle, A being the highest and BC horizontal. A string whose length equals four times a side of the triangle is hung over them, and its two ends attached to a weight W. Find the pressure on each peg.

17. A sphere, of weight W, is hung from a point in a wall by a string of length l attached to its centre. If T is the tension of the string, R the pressure on the wall, r the radius of the sphere, shew that
$$R : T : W = r : l : (l^2 - r^2)^{\frac{1}{2}}.$$

18. Two inclined planes of equal bases, b, and altitudes, h, are placed facing each other on a smooth table; a sphere of weight W is placed between them, and they are prevented from sliding apart by a string. Shew that the tension of the string $= \frac{1}{2} W \cdot \frac{h}{b}$.

19. A line of telegraph posts is erected along a road, and a wire is carried along the tops in the usual manner. At a certain point, where a post stands, the road makes a turn through 60°. To keep the post in its vertical position one extremity of a wire rope is attached to the middle point of the post, and the other to a point in the ground, midway between the two directions of the road. If the rope make an angle of 30° with the post, and the telegraph wire be supposed horizontal, shew that the tension of the rope is 4 times the tension of the wire.

20. Prove the following construction for finding the centre of gravity of a quadrilateral. E is the point of intersection of the diagonals, F and G are points in the diagonals, dividing them into segments equal to those, into which they are divided at E. The centre of gravity of the quadrilateral coincides with the centre of gravity of the triangle EFG.

21. Three smooth pegs are fastened in a vertical plane, so as to form an isosceles triangle with its base horizontal, vertex downwards, and vertical angle equal to 120°. A fine string, with a weight W attached to each extremity, is passed under the lower peg and over the other two. Find the pressure on each peg. Find also the vertical pressure on each peg, and shew how the whole weight is supported.

22. Six forces, of 1, 2, 3, 4, 5 and 6 lbs., act from the centre, O, of a regular hexagon towards the angular points, A, B, C, D, E, F, respectively. Shew that the direction of their resultant is OE; and its magnitude equals 6 lbs.

23. A heavy rod, AB, is moveable about a fulcrum at C, and is kept in a horizontal position by a string which has its ends attached to A and B, and passes over two fixed pullies, D, E, which are respectively vertically above A and B. Shew that wherever C is, the tension of the string equals half the weight of the rod.

24. If triangles be described upon a given base AB, and the sides be taken to represent forces, two of which tend from A, and the third towards B, shew that the resultant of these forces is constant in direction and magnitude.

25. AD, BE, CF are the perpendiculars drawn from the angular points of a triangle ABC to the opposite sides. Forces at D act in the directions DE, DF; find their relative magnitudes if they be kept in equilibrium by a force in the direction AD.

26. The two arms of a lever are equal, and they form two sides of a square, the fulcrum being at their intersection. Two equal forces act at their extremities along the remaining sides: find the pressure on the fulcrum.

ANSWERS TO EXAMPLES.

I.　(p. 25.)

1. 12 lbs., or 6 lbs., or 4 lbs., or 2 lbs.
2. 29 lbs., or 19 lbs., or 7 lbs., or 3 lbs.
3. 150 lbs.
4. 60 lbs.
7. 5 lbs., and $5\sqrt{3}$ lbs.
8. $\sqrt{34 + 15\sqrt{3}}$ lbs.
9. $\sqrt{219}$ lbs.
10. $\sqrt{202 - 99\sqrt{2}}$ lbs.
11. $\sqrt{41 + 20\sqrt{2}}$ lbs.
12. $\sqrt{3}$ times either of the equal forces.
14. $\dfrac{10}{3}\sqrt{3}$ lbs.

II.　(p. 31.)

2. The forces are equal.
4. If F be one of the equal forces, third force $= \sqrt{3} \cdot F$.
6. $\sqrt{19}$ lbs.
7. $(\sqrt{6}-1)$ lbs.
8. Direction, North.　Magnitude, P.
9. 90°, 150°, 180°.
10. $P : Q : R = 1 : 1 : \sqrt{2}$.
12. $1 : 2 : \sqrt{3}$.
13. $\sqrt{3} \cdot P$.
14. It makes an angle of 60° with the smaller.
15. $\sqrt{2} : 1$.
16. $1 : \sqrt{3}$.

17. (1) Resultant makes an angle of 30° with greater.

(2) Resultant makes an angle of 67½° with force that acts East.   22. 120°.   23. The forces must be opposite, two and two.   25. R is equal and opposite R'.
26. X lies in EA produced so that $AX = AE$.

III. (p. 40.)

1. $4\sqrt{3}$ lbs. and $8\sqrt{3}$ lbs.   2. (1) $\dfrac{10\sqrt{3}}{3}$ lbs.,

(2) 10 lbs.   3. $-\dfrac{1}{3}$.   5. 24 lbs.

8. $1 : \sqrt{3}$.   9. $\dfrac{7}{8}$ and $\dfrac{11}{16}$.

11. The radius through P's position makes with the vertical an angle whose tangent is $\dfrac{Q}{P}$.

13. Resultant $= 2\sqrt{(47 + 16\sqrt{3})}$, and makes with OA an angle whose tangent $= \dfrac{17 + \sqrt{3}}{22}$.

IV. (p. 45.)

1. $3\frac{1}{3}$ inches.   2. 4 inches.   3. 4 ft. 10 in.
4. 2 ft. 6 in.   5. 15 lbs.

V. (p. 52.)

1. $3\frac{1}{4}$ ft. and $2\frac{1}{4}$ ft.   2. $3 : 1$.   3. 63 lbs. and 36 lbs.
4. 3 ft.   5. 21 inches from the smaller weight.
6. $5 : 2$.   7. $P - Q$ or $Q - P$.   8. 9 inches from the larger weight. If the forces were *parallel* they might act in opposite directions, and then the fulcrum would not be between the points of application.   9. 6 ft. from the smaller weight.   10. 8 inches nearer to the smaller weight.   11. $104\frac{1}{4}$ lbs.   12. 2 ft.   13. $20\frac{4}{25}$ inches.
14. 12 lbs.   15. (1) $\dfrac{8\sqrt{3}}{5}$ lbs.   (2) $1\frac{1}{4}$ lbs.   (3) $\dfrac{6\sqrt{2}}{5}$ lbs.

VI. (p. 65.)

1. 16 inches from end to which the 1 lb. is attached.
2. $3\frac{4}{7}$ inches from the 4 lbs.   3. $3\frac{1}{2}$ inches from the

fourth weight.   4. $\frac{26}{31}$ ft. from the weight 16.

10. Let D be the centre of gravity of A and B, and E that of A and C. Join CD, BE, and let them intersect in O. AO produced will cut BC in the point required.

13. To keep the vertical line of the common centre of gravity of himself and the weight within his base.

14. 11 inches.   15. Midway between the middle points of AB and CD.   17. $\sqrt{3}$ times the side of the square.

18. $\sqrt{2} : 1$.   19. If O be the centre of gravity of the triangle, and the 4 lbs. be at B, and the 5 lbs. at C, the centre of gravity required is one-fourth of the distance of O from the point of trisection of BC nearest C.

20. Draw from the angular point, at which the smallest weight acts, a line to the middle point of the opposite side. The centre of gravity is four-fifths of this line from the angular point.   23. $45^\circ + A$.   24. 30°.   26. At C.

27. If G be the centre of gravity, $AG : AD = 4 : 9$.

28. 12 inches.   29. If G be the centre of gravity, G divides OC, so that $OG = \frac{OC}{3}$.   32. The middle point of the line joining the centres of gravity of the equal triangles.   33. If $2a$ be the side of the square, the vertex of the triangle is at a distance $2a - a\sqrt{3}$ from the centre of the square.   35. The centre of gravity is on the line joining the vertex to the middle point of the base, and at a distance from the vertex $= \frac{7}{9}$ ths distance of the middle point.

36. 2 inches.   37. $\sqrt{3}$ (side of square).   38. If $2a$ be a side of the square, O the centre of gravity of the square, E that of the triangle, and G that of both, $OG = \frac{a(\sqrt{3}+1)}{4+\sqrt{3}}$.

39. It divides the line joining the middle points of the parallel sides into parts as $4 : 5$, and it is nearer to the greater side.   40. (1) The point in which the diagonals intersect.   (2) It divides the line joining the centres of 9 and 1 in the ratio $5 : 4$.   41. $\frac{14}{27}$ of side of square from GA and $\frac{4}{9}$ of side of square from GE.

42. $\frac{11}{30}$ of GC from C.   44. Let a = length of line

between the middle point of the omitted side and its opposite; then the centre of gravity is on this line and at a distance $\frac{a}{10}$ from the centre of the hexagon.   45. One-third.

46. Draw a radius *CD* from any point, and place the 5 lbs. at *C*. Produce *CD* to *E*, so that $ED = \frac{5}{6} . CD$. Draw the chord *AEB* at right angles to *CE*, and place the weights of 3 lbs. at *A* and *B*.   47. Let the straight line cut the line joining the vertex to the middle point of the base into two parts *m* and *n*, *m* being nearer the vertex. Then the centre of gravity is on this latter line at a distance from the vertex

$$= \frac{2}{3} \frac{\{(m+n)^3 - m^3\}}{\{(m+n)^2 - m^2\}}.$$

49. The centre of gravity bisects the line joining 3*w* with the point of trisection of the opposite side nearest to 2*w*, and it divides the line from *w* to the opposite side into parts as 5 : 1; and that from 2*w* to the opposite side into parts as 2 : 1.

50. 2 lbs.   51. Weight at *B* : weight at $C = AC^2 : AB^2$.

53. At the middle point of the hypotenuse of the triangle.

VII. (p. 87.)

1. 3 ft. from *A* ; 10 lbs.   2. $20\frac{1}{4}$ lbs. and $15\frac{3}{4}$ lbs.

3. $3\frac{3}{7}$ lbs.   4. 1 inch from *A*; 10 lbs.   5. $\frac{1}{8}$ and $\frac{2}{3}$.

7. 30 inches from the end where 10 lbs. hangs ; 60 lbs.

8. 25 lbs. 9. 3 lbs. and $3\frac{1}{3}$ lbs. 10. $2\frac{1}{2}$ lbs. and $5\frac{1}{2}$ lbs.; 6 lbs.

11. 1 ft. from weight of 3 lbs.   12. Increased.   13. 6 ft.

14. 1 inch from fulcrum.   15. 15 lbs.; at advantage.

16. $10\frac{2}{3}$ st. and $9\frac{1}{3}$ st.   17. $\frac{2(ra-b)}{r-1}$.   18. The point of junction.   19. $5\frac{11}{17}$ ft. from the 25 lbs.   20. $\frac{1}{4}$ of length of rod from one prop.   21. 56 lbs. and 96 lbs.   22. 1280 lbs.

23. $9\frac{37}{38}$ ft. from that end.   24. (i) At a distance from the given point $= n-1$ times the distance of *P* from it. (ii) 11.

25. 11 lbs. at end where 3 lbs. acts.   26. A weight 8 times that of the shorter rod.   27. 14 lbs. ; 12 inches.

28. 24 ft.; 1 lb.   29. $28\frac{4}{5}$ lbs.   30. 38 ft. from fulcrum on first arm.   31. $7\frac{1}{2}$ lbs.   32. $\frac{16\sqrt{3}}{3}$ feet.   33. One way is to place the upper rod so that its 1 lb. weight lies on the 9 lbs. weight of the lower, and the rest of the upper on the

rest of the lower. Another way is to place the upper with the
1 lb. weight projecting $1\frac{2}{3}$ ft. beyond the 1 lb. weight of the
lower.   34.   Inclined at 30° to the lever; $2\sqrt{3}$ lbs.

35. $\frac{3}{4}$ (weight of shorter beam).   38.   1 : 2   39. $\dfrac{16\sqrt{6}}{3}$ lbs.

VIII.   (p. 96.)

1. $42\frac{2}{5}$ lbs.   4.   At the point of support.   6.   40 lbs.
7. $18\frac{2}{7}$ oz.   8.   2 : 3.   9.   4 shillings.   10.   Taking
the diagram on p. 95, the graduations from O towards A will

run thus : the first will be $\dfrac{1}{16n+1}$ the length of AO, the second

$\dfrac{2\,AO}{16n+2}$, the third $\dfrac{3\,AO}{16n+3}$ ... and so on.

IX.   (p. 99.)

1.   6 in.   2.   1 lb.   3.   4 lbs.   4.   15 cwt.   5.   At
right angles to the direction of the weight.   6.   When the
direction of P is a tangent to the axle and on the opposite side
to W.   7.   108 lbs.   9. $\sqrt{2}$: 1.

X.   (p. 105.)

1.   120°.   3.   1 : $\sqrt{3}$.   4.   1 lb.   5.   8 lbs.   6. $3\frac{1}{4}$ lbs.
7.   12 lbs.   8.   2 lbs.   10.   1 lb.   11. W.   15.   It
divides the distance between the points where the strings are
fastened to the beam in the ratio 1 : 2, being nearer to the
string passing under the lower pulley.   16.   20 lbs.
18. $\frac{13}{15}$ of his weight.   19.   3 times the power.   20.   2 lbs.
22.   If $2a$ be the length of the bar, the weight must be placed
$\dfrac{4a}{7}$ from third string.   23.   1 : 57

XI.   (p. 112.)

1. $1\frac{3}{13}$ tons.   2.   8 lbs.   3. $R = \dfrac{3W}{5}$, $P = \dfrac{4W}{5}$.

4.   9 lbs.   5.   3 lbs.; 30°.   6.   12 lbs.   7. $1\frac{200}{333}$ tons.
8.   60°.   9.   2 lbs.   10.   Any force greater than 56 lbs.

11. $12\sqrt{3}$ lbs.   12. $\dfrac{10\sqrt{3}}{3}$ lbs.   13. 60°.   14. $\dfrac{2-\sqrt{2}}{2}$
times the weight.   15. 2 lbs.   17. The power acting
parallel to the plane.   18.   As soon as the inclination is
greater than 30°.   19. If l and l' be the lengths of the
planes, $W : W' = l : l'$.   20. 60°.   21. 60°.   23. 30°.
24. $3\frac{3}{4}$ lbs.   25. $W . \tan a - W . \sin a$.

MISCELLANEOUS EXERCISES (p. 121).

1. No.   5. $P_1 : P_2 : W = 3 : 1 : \sqrt{10}$.   8. 15 lbs. and 20 lbs.
9. Tension of each string $= \dfrac{W\sqrt{3}}{3}$.   12. $\sqrt{6} : \sqrt{3}+1 : 2$.

16. Pressure on $A = W$. Pressure on each of B and $C = \dfrac{W\sqrt{3}}{3}$.

21. Pressure on each upper peg $= W\sqrt{3}$. Pressure on lower
peg $= W$. Vertical pressure on each upper peg $= \dfrac{3W}{2}$; on
lower peg $= W$. Thus the whole weight supported, *i.e.*, $2W$,
is so distributed as to produce a downward pressure on each of
the upper pegs $= \dfrac{3}{2}W$, and an upward pressure on the lower
peg $= W$.   24. Resultant $= 2AB$.   26. Pressure $= \sqrt{2}. P$,
where P represents each force.

MUIR, PATERSON AND BRODIE, PRINTERS, EDINBURGH.

www.ingramcontent.com/pod-product-compliance
Lightning Source LLC
Chambersburg PA
CBHW021933190326
41519CB00009B/1009